W. R. Fahrner (Editor)

Nanotechnology and Nanoelectronics
Materials, Devices, Measurement Techniques

W. R. Fahrner (Editor)

Nanotechnology and Nanoelectronics

Materials, Devices, Measurement Techniques

With 218 Figures

 Springer

Prof. Dr. W. R. Fahrner
University of Hagen
Chair of Electronic Devices
58084 Hagen
Germany

Library of Congress Control Number: 2004109048

ISBN 3-540-22452-1 Springer Berlin Heidelberg New York

Springer is a part of Springer Science + Business Media GmbH

springeronline.com

© Springer-Verlag Berlin Heidelberg 2005
Printed in Germany

Typesetting: Digital data supplied by editor
Cover-Design: medionet AG, Berlin
Printed on acid-free paper   62/3111 RW   5 4 3 2 1   SPIN 11577942

Preface

*Split a human hair thirty thousand times, and
you have the equivalent of a nanometer.*

The aim of this work is to provide an introduction into nanotechnology for the scientifically interested. However, such an enterprise requires a balance between comprehensibility and scientific accuracy. In case of doubt, preference is given to the latter.

Much more than in microtechnology – whose fundamentals we assume to be known – a certain range of engineering and natural sciences are interwoven in nanotechnology. For instance, newly developed tools from mechanical engineering are essential in the production of nanoelectronic structures. Vice versa, mechanical shifts in the nanometer range demand piezoelectric-operated actuators. Therefore, special attention is given to a comprehensive presentation of the matter. In our time, it is no longer sufficient to simply explain how an electronic device operates; the materials and procedures used for its production and the measuring instruments used for its characterization are equally important.

The main chapters as well as several important sections in this book end in an evaluation of future prospects. Unfortunately, this way of separating coherent description from reflection and speculation could not be strictly maintained. Sometimes, the complete description of a device calls for discussion of its inherent potential; the hasty reader in search of the general perspective is therefore advised to study this work's technical chapters as well.

Most of the contributing authors are involved in the "Nanotechnology Cooperation NRW" and would like to thank all of the members of the cooperation as well as those of the participating departments who helped with the preparation of this work. They are also grateful to Dr. H. Gabor, Dr. J. A. Weima, and Mrs. K. Meusinger for scientific contributions, fruitful discussions, technical assistance, and drawings. Furthermore, I am obliged to my son Andreas and my daughter Stefanie, whose help was essential in editing this book.

Hagen, May 2004   W. R. Fahrner

Contents

Contributors

Prof. Dr. rer. nat. Wolfgang R. Fahrner (Editor)
University of Hagen
Haldenerstr. 182, 58084 Hagen, Germany

Prof. Dr.-Ing. Ulrich Hilleringmann
University of Paderborn
Warburger Str. 100, 33098 Paderborn, Germany

Dr.-Ing. John T. Horstmann
University of Dortmund
Emil-Figge-Str. 68, 44227 Dortmund, Germany

Dr. rer. nat. habil. Reinhart Job
University of Hagen
Haldenerstr. 182, 58084 Hagen, Germany

Prof. Dr.-Ing. Heinz-Christoph Neitzert
University of Salerno
Via Ponte Don Melillo 1, 84084 Fisciano (SA), Italy

Prof. Dr.-Ing. Hella-Christin Scheer
University of Wuppertal
Rainer-Gruenter-Str. 21, 42119 Wuppertal, Germany

Dr. Alexander Ulyashin
University of Hagen
Haldenerstr. 182, 58084 Hagen, Germany

Prof. Dr. rer. nat. Andreas Dirk Wieck
University of Bochum
Universitätsstr. 150, NB03/58, 44780 Bochum, Germany

Abbreviations

AES	Auger electron spectroscopy
AFM	Atomic force microscope / microscopy
ASIC	Application-specific integrated circuit
BSF	Back surface field
BZ	Brillouin zone
CARL	Chemically amplified resist lithography
CCD	Charge-coupled device
CMOS	Complementary metal–oxide–semiconductor
CNT	Carbon nanotube
CVD	Chemical vapor deposition
CW	Continuous wave
Cz	Czochralski
DBQW	Double-barrier quantum-well
DFB	Distributed feedback (QCL)
DLTS	Deep level transient spectroscopy
DOF	Depth of focus
DRAM	Dynamic random access memory
DUV	Deep ultraviolet
EBIC	Electron beam induced current
ECL	Emitter-coupled logic
ECR	Electron cyclotron resonance (CVD, plasma etching)
EDP	Ethylene diamine / pyrocatechol
EEPROM	Electrically erasable programmable read-only memory
EL	Electroluminescence
ESR	Electron spin resonance
ESTOR	Electrostatic data storage
Et	Ethyl
EUV	Extreme ultraviolet
EUVL	Extreme ultraviolet lithography
EXAFS	Extended x-ray absorption fine-structure studies
FEA	Field emitter cathode array
FET	Field effect transistor
FIB	Focused ion beam
FP	Fabry-Perot
FTIR	Fourier transform infrared
FWHM	Full width at half maximum

HBT	Hetero bipolar transistor
HEL	Hot-embossing lithography
HEMT	High electron mobility transistor
HIT	Heterojunction with intrinsic thin layer
HOMO	Highest occupied molecular orbital
HREM	High resolution electron microscope / microscopy
IC	Integrated circuit
ICP	Inductively coupled plasma
IMPATT	Impact ionization avalanche transit time
IPG	In plane gate
IR	Infrared
ITO	Indium–tin–oxide
ITRS	International technology roadmap for semiconductors
Laser	Light amplification by stimulated emission of radiation
LBIC	Light beam induced current
LDD	Lightly doped drain
LED	Light-emitting diode
LEED	Low energy electron diffraction
LMIS	Liquid metal ion source
LPE	Liquid phase epitaxy
LSS	Lindhardt, Scharff, Schiøtt (Researchers)
LUMO	Lowest unoccupied molecular orbital
M	Metal
MAL	Mould-assisted lithography
MBE	Molecular beam epitaxy
μCP	Microcontact printing
MCT	Mercury cadmium telluride
Me	Methyl
MEMS	Micro electro-mechanical system
MIS	Metal–insulator–semiconductor
MMIC	Monolithic microwave integrated circuit
MOCVD	Metallo-organic chemical vapor deposition
MODFET	Modulation-doped field-effect transistor
MOLCAO	Molecular orbitals as linear combinations of atomic orbitals
MOS	Metal–oxide–semiconductor
MOSFET	Metal–oxide–semiconductor field effect transistor
MPU	Microprocessor unit
MQW	Multi quantum well
MWNT	Multi wall nanotubes
NA	Numerical aperture
NAND	Not and
NDR	Negative differential resistance
Nd:YAG	Neodymium yttrium aluminum garnet (laser)
NIL	Nanoimprint lithography
NMOS	n-Channel metal–oxide–semiconductor (transistor)
NMR	Nuclear magnetic resonance
NOR	Not or

PADOX	Pattern-dependent oxidation
PDMS	Polydimethylsiloxane
PE	Plasma etching
PECVD	Plasma-enhanced chemical vapor deposition
PET	Polyethyleneterephthalate
PL	Photoluminescence
PLAD	Plasma doped
PMMA	Polymethylmethacrylate
PREVAIL	Projection reduction exposure with variable axis immersion lenses
PTFE	Polytetrafluorethylene (Teflon®)
PVD	Physical vapor deposition
QCL	Quantum cascade laser
QSE	Quantum size effect
QWIP	Quantum well infrared photodetector
RAM	Random access memory
RBS	Rutherford backscattering spectrometry
RCA	Radio Corporation of America (Company)
RF	Radio frequency
RHEED	Reflection high-energy electron diffraction
RIE	Reactive ion etching
RITD	Resonant interband tunneling diode
RTA	Rapid thermal annealing
RTBT	Resonant tunneling bipolar transistor
RTD	Resonant tunneling diode
SAM	Self-assembling monolayer
SCALPEL	Scattering with angular limitation projection electron beam lithography
SCZ	Space charge zone
SEM	Scanning electron microscopy
SET	Single electron transistor
SFIL	Step and flash imprint lithography
SHT	Single hole transistor
SIA	Semiconductor Industry Association
SIMOX	Separation by implantation of oxygen
SIMS	Secondary ion mass spectroscopy
SL	Superlattice
SMD	Surface-mounted device
SOI	Silicon on insulator
SOS	Silicon on sapphire
STM	Scanning tunneling microscope / microscopy
SWNT	Single wall nanotubes
TA	Thermal analysis
TED	Transferred electron device
TEM	Transmission electron microscopy
TEOS	Tetraethylorthosilicate
TFT	Thin film transistor
TMAH	Tetramethylammonium hydroxide
TSI	Top surface imaging

TTL	Transistor-transistor logic
TUBEFET	Single carbon nanotube field-effect transistor
UHV	Ultrahigh vacuum
ULSI	Ultra large scale integration
UV	Ultraviolet
VHF	Very high frequency (30–300 MHz; 10–1 m)
VLSI	Very large scale integration
VMT	Velocity-modulated transistor
V-PADOX	Vertical pattern-dependent oxidation
VPE	Vapor phase epitaxy
XOR	Exclusive or
XRD	X-ray diffraction
ZME	Zeolite modified electrode

1 Historical Development

1.1 Miniaturization of Electrical and Electronic Devices

At present, development in electronic devices means a race for a constant decrease in the order of dimension. The general public is well aware of the fact that we live in the age of microelectronics, an expression which is derived from the size (1 µm) of a device's active zone, e.g., the channel length of a field effect transistor or the thickness of a gate dielectric. However, there are convincing indications that we are entering another era, namely the age of nanotechnology. The expression "nanotechnology" is again derived from the typical geometrical dimension of an electronic device, which is the nanometer and which is one billionth (10^{-9}) of a meter. 30,000 nm are approximately equal to the thickness of a human hair. It is worthwhile comparing this figure with those of early electrical machines, such as a motor or a telephone with their typical dimensions of 10 cm. An example of this development is given in Fig. 1.1.

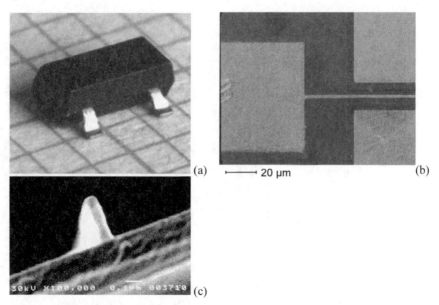

(a) ⊢——— 20 µm   (b)

30kU X100,000 0.1µm 003710 (c)

Fig. 1.1   (a) Centimeter device (SMD capacity), (b) micrometer device (transistor in an IC), and (c) nanometer device (MOS single transistor)

1.2   Moore's Law and the SIA Roadmap

From the industrial point of view, it is of great interest to know which geometrical dimension can be expected in a given year, but the answer does not only concern manufacturers of process equipment. In reality, these dimensions affect almost all electrical parameters like amplification, transconductance, frequency limits, power consumption, leakage currents, etc. In fact, these data have a great effect even on the consumer. At first glance, this appears to be an impossible prediction of the future. However, when collecting these data from the past and extrapolating them into the future we find a dependency as shown in Fig. 1.2. This observation was first made by Moore in 1965, and is hence known as Moore's law.

A typical electronic device of the fifties was a single device with a dimension of 1 cm, while the age of microelectronics began in the eighties. Based on this figure, it seems encouraging to extrapolate the graph, for instance, in the year 2030 in which the nanometer era is to be expected. This investigation was further pursued by the Semiconductor Industry Association (SIA) [1]. As a result of the above-mentioned ideas, predictions about the development of several device parameters have been published. A typical result is shown in Table 1.1.

These predictions are not restricted to nanoelectronics alone but can also be valid for materials, methods, and systems. There are schools and institutions which are engaged in predictions of how nanotechnology will influence or even rule our lives [2]. Scenarios about acquisition of solar energy, a cure for cancer, soil detoxification, extraterrestrial contact, and genetic technology are introduced. It should be considered, though, that the basic knowledge of this second method of prediction is very limited.

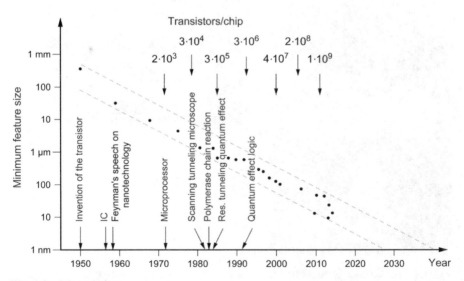

Fig. 1.2   Moore's law

Table 1.1   Selected roadmap milestones

Year	1997	1999	2001	2003	2006	2009	2012
Dense lines, nm	250	180	150	130	100	70	50
Iso. lines (MPU gates), nm	200	140	120	100	70	50	35
DRAM memory (introduced)	267 M	1.07 G	[1.7 G]	4.29 G	17.2 G	68.7 G	275 G
MPU: transistors per chip	11 M	21 M	40 M	76 M	200 M	520 M	1.4 G
Frequency, MHz	750	1200	1400	1600	2000	2500	3000
Minimum supply voltage V_{dd}, V	1.8– 2.5	1.5– 1.8	1.2– 1.5	1.2– 1.5	0.9– 1.2	0.6– 0.9	0.5– 0.6
Max. wafer diameter, mm	200	300	300	300	300	450	450
DRAM chip size, mm^2 (introduced)	280	400	445	560	790	1120	1580
Lithography field size, mm^2	22·22 484	25·32 800	25·34 850	25·36 900	25·40 1000	25·44 1100	25·52 1300
Maximum wiring levels	6	6–7	7	7	7–8	8–9	9
Maximum mask levels	22	22–24	23	24	24–26	26–28	28
Density of electrical DRAM defects (introduced), 1/m^2	2080	1455	[1310]	1040	735	520	370

MPU: microprocessor unit, DRAM: dynamic random access memory

2   Quantum Mechanical Aspects

2.1   General Considerations

Physics is the classical material science which covers two extremes: on the one hand, there is atomic or molecular physics. This system consists of one or several atoms. Because of this limited number, we are dealing with sharply defined discrete energy levels. On the other side there is solid-state physics. The assumption of an infinitely extended body with high translation symmetry also makes it open to mathematical treatment. The production of clusters (molecules with 10 to 10,000 atoms) opens a new field of physics, namely the observation of a transition between both extremes. Of course, any experimental investigation must be followed by quantum mechanical descriptions which in turn demand new tools.

Another application of quantum mechanics is the determination of stable molecules. The advance of nanotechnology raises hopes of constructing mechanical tools within human veins or organs for instance, valves, separation units, ion exchangers, molecular repair cells and depots for medication. A special aspect of medication depot is that both the container and the medicament itself would have to be nanosynthesized.

Quantum mechanics also plays a role when the geometrical dimension is equal to or smaller than a characteristic wavelength, either the wavelength of an external radiation or the de Broglie wavelength of a particle in a bound system. An example of the first case is diffraction and for the second case, the development of discrete energy levels in a MOS inversion channel.

2.2   Simulation of the Properties of Molecular Clusters

One of the first theoretical approaches to nanotechnology has been the simulated synthesis of clusters (molecular bonding of ten to some ten thousand atoms of different elements). This approach dates back to the 1970s. In a simulation, a Hamilton operator needs to be set up. In order to do so, some reasonable arrangement of the positions of the atoms is selected prior to the simulation's beginning. An adiabatic approach is made for the solution of the eigenvalues and eigenfunctions. In our case, this means that the electronic movement is much faster than that of the atoms. This is why the electronic system can be separated from that of the atoms and leads to an independent mathematical treatment of both systems. Because of the electronic system's considerably higher energy, the Schrödinger equation for the electrons can be calculated as a one-electron solution. The method used for the calculation is called MOLCAO *(molecular orbitals as linear combinations of atomic orbitals)*. As can be derived from the acronym, a molecular orbital is assumed to be a linear combination of orbitals from the atomic component as is known from the theory of single atoms. The eigenvalues and coefficients are determined by diagonalization in accordance with the method of linear algebra. Then the levels, i.e., the calculated eigenenergies will be filled with electrons according to the Pauli principle. Thereafter the total energy can be calculated by multiplying the sum of the eigenenergies by the electrons in these levels. A variation calculation is performed at the end in order to obtain the minimum energy of the system. The parameter to be varied is the geometry of the atom, i.e., its bonding length and angle. The simulations are verified by application on several known properties of molecules (such as methane and silane), carbon-containing clusters (like fullerenes) and vacancy-containing clusters in silicon. This method is not only capable of predicting new stable clusters but is also more accurate in terms of delivering their geometry, energy states, and optical transitions. This is already state-of-the-art [3–5]. Thus, no examples are given.

Starting from here, a great number of simulations are being performed for industrial application like hydrogen storage in the economics of energy, the synthesis of medication in the field of medicine or the development of lubricants for automobiles. As an example, we will consider the interaction between hydrogen atoms and fullerenes (Fig. 2.1). An incomplete fullerene (a fullerene with a vacancy) is selected. If placed in a hydrogen environment (14 in the simulation), the aforesaid vacancy captures four hydrogen atoms. In conclusion, a vacancy can take at least four hydrogen atoms. It is simple to produce fullerenes with a higher number of vacancies so that a fullerene can eventually be expected to be an active

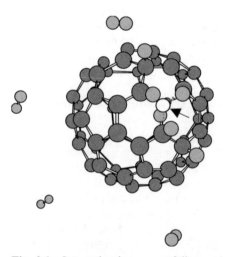

Fig. 2.1   Interaction between a fullerene (which contains a vacancy) and hydrogen. The dark-gray circles represent carbon, the light-gray ones hydrogen, and the empty circle (arrow) represents a vacancy with dangling bonds.

storage medium for hydrogen (please consider the fact that the investigations are not yet completed). There are two further hydrogen atoms close to the vacancy which are weakly bonded to the hydrogen atoms but can also be part of the carbon-hydrogen complex.

A good number of commercial programs are available for the above-named calculations. Among these are the codes Mopac, Hyperchem, Gaussian, and Gamess, to name but a few. All of these programs require high quality computers. The selection of 14 interactive hydrogen atoms was done in view of the fact that the calculation time be kept within a reasonable limit.

In other applications mechanical parts such as gears, valves, and filters are constructed by means of simulation (Fig. 2.2). These filters are meant to be employed in human veins in order to separate healthy cells from infected ones (e.g., by viruses or bacteria). Some scientists are even dreaming of replacing the passive filters by active machines (immune machines) which are capable of detecting penetrating viruses, bacteria and other intruders. Another assignment would be the reconstruction of damaged tissues and even the replacement of organs and bones.

Moreover, scientists consider the self-replicating generation of the passive and active components discussed above. The combination of self-replication and medicine (especially when involving genetic engineering) opens up a further field of possibilities but at the same time provokes discussions about seriousness and objectives.

2.3   Formation of the Energy Gap

As discovered above, clusters are found somewhere in the middle between the single atom on one side and the infinitely extended solid state on the other. Therefore, it should be possible to observe the transition from discrete energy states to the energy gap of the infinitely extended solid state on the other side. The results of such calculations are presented in Figs. 2.3 and 2.4.

Note that the C_5H_{12} configuration in Fig. 2.3 is *not* the neopentane molecule (2,2-dimethylpropane). It is much more a C_5 arrangement of five C atoms as near-

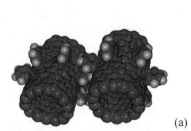

(a)   (b)

Fig. 2.2   (a) Nanogear [6], (b) nanotube or nanofilter [7]

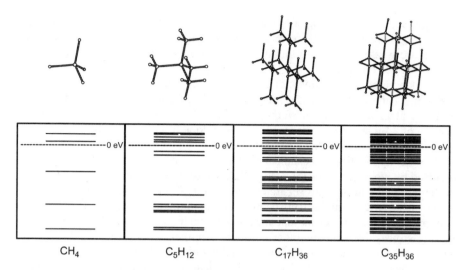

Fig. 2.3   Development of the diamond band gap

est neighbors which are cut out of the diamond. For the purpose of electronic satu-ration 12 hydrogen atoms are hung on this complex. The difference to a neopen-tane molecule lies in the binding lengths and angles.

In the examples concerning carbon and silicon, the development of the band structure is clearly visible. In another approach the band gap of silicon is deter-mined as a function of a typical length coordinate, say the cluster radius or the length of a wire or a disc. In Fig. 2.5, the band gap versus the reciprocal of the length is shown [8]. For a solid state, the band gap converges to its well known value of 1.12 eV.

It is worthwhile comparing the above-mentioned predictions with subsequent experimental results [9]. The band gap of Si_n clusters is investigated by photo-electron spectroscopy. Contrary to expectations, it is shown that almost all clusters from $n = 4$ to 35 have band gaps smaller than that of crystalline silicon (see Fig. 2.6). These observations are due to pair formation and surface reconstruction.

Scientists are in fact interested in obtaining details which are even more specific. For example, optical properties are not only determined through the band gap but through the specific dependency of the energy bands on the wave vectors. It is a much harder theoretical and computational assignment to determine this dependency. An earlier result [10] for SiC cluster is reproduced in Fig. 2.7.

2.4   Preliminary Considerations for Lithography

An obvious effect of the quantum mechanics on the nanostructuring can be found in lithography. For readers with little experience, the lithographic method will be briefly explained with the help of Fig. 2.8.

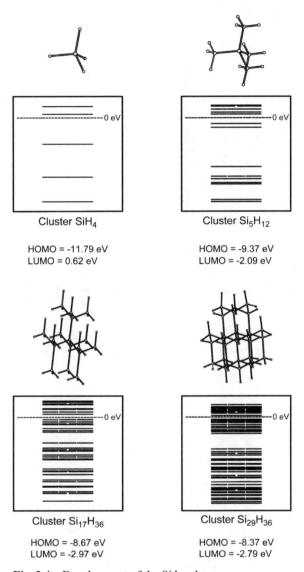

Fig. 2.4   Development of the Si band gap

A wafer is covered with a photoresist and a mask containing black/transparent structures is laid on top of it. If the mask is radiated with UV light, the light will be absorbed in the black areas and transmitted in the other positions. The UV light subsequently hardens the photoresist under the transparent areas so that it cannot be attacked by a chemical solution (the developer). Thus, a window is opened in the photoresist at a position in the wafer where, for instance, ion implantation will be performed. The hardened photoresist acts as a mask which protects those areas that are not intended for implantation.

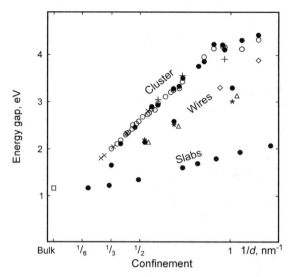

Fig. 2.5   Energy gaps *vs.* confinement. The different symbols refer to different computer programs which were used in the simulation.

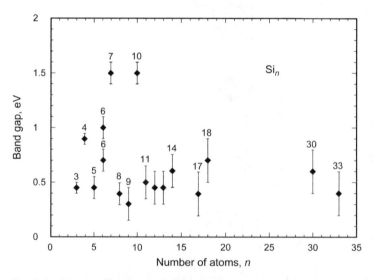

Fig. 2.6   Measured band gaps for silicon clusters

Up to now, a geometrical light path has been tacitly assumed i.e., an exact reproduction of the illuminated areas. However, wave optics teaches us that this not true [11]. The main problem is with the reproduction of the edges. From geometrical optics, we expect a sharp rise in intensity from 0 % (shaded area) to 100 % (the irradiated area). The real transition is shown in Fig. 2.9.

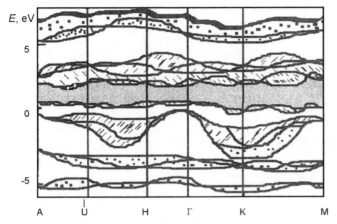

Fig. 2.7   *E-k* diagram for nanocrystalline SiC

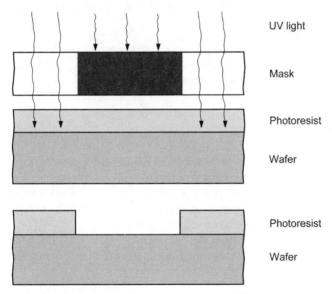

Fig. 2.8   (Optical) lithography

It turns out that the resolution of an image produced cannot be better than approximately one wavelength of the light used. In this context, "light" means anything that can be described by a wavelength. This includes x-rays, synchrotron radiation, electrons and ions. As an example, the wavelength of an incident electron is given by

$$\lambda = \frac{h}{\sqrt{2q \, V \, m_e}}$$ (2.1)

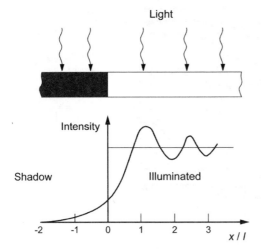

Fig. 2.9   Diffraction image of a black/transparent edge. l is a length which is equivalent to the wavelength of the incident light.

(h is the Planck constant, m_e the mass of electron, q the elementary charge, V the accelerating voltage). The different types of lithography, their pros and cons, and their future prospects will be discussed in the section about nanoprocessing.

2.5   Confinement Effects

In the early days of quantum mechanics, one considered the case of a particle, e.g., an electron that is confined in a tightly bounded potential well V with high walls. It is shown that within the walls ($0 < x < a$), the wave function of the electron is oscillatory (a standing wave) while it presents an exponential decaying function in the forbidden zone outside the walls ($x < 0, x > a$), Fig. 2.10.

Thus, the particle's behavior departs from the rule in two respects: (i) Discrete energy levels E_i and wave functions are obtained as a result of the demand for

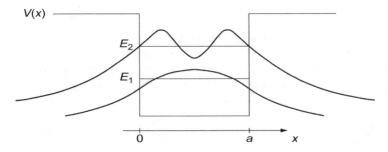

Fig. 2.10   Particles in a potential well

continuity and continuous differentiation of the wave function on the walls [12]. This is contrary to classical macroscopic findings that the electron should be free to accept all energies between the bottom and the top margins of the potential well. (ii) The particle shows a non-vanishing probability that it moves outside the highly confined walls. In particular, it has the chance to penetrate a neighboring potential well with high walls. In such a case, we are dealing with the possibility of so-called tunneling.

In anticipation, both consequences will be briefly shown with the help of examples. A detailed description will be given in the sections dedicated to nanodevices.

2.5.1 Discreteness of Energy Levels

The manufacturing of sufficiently closely packed potential wells in an effort to investigate the above-mentioned predictions has not been easy. Mostly they are investigated with the help of electrons which are bound to crystal defects, e.g., by color centers. Meanwhile, a good number of experimental systems via which quantization occurs are available. One example is the MOS varactor. Let us assume that it is built from a p-type wafer. We will examine the case in which it is operated in inversion. The resulting potential for electrons and the wave functions are schematically presented in Fig. 2.11.

The normal operation of a MOS transistor is characterized by the electrons being driven from the source to the drain, i.e., perpendicular to plane of the figure. Ideally, they can only move within these quantum states (the real behavior is modified through phonon interaction). The continuation of this basic assumption leads to a way with which the fine-structure constant

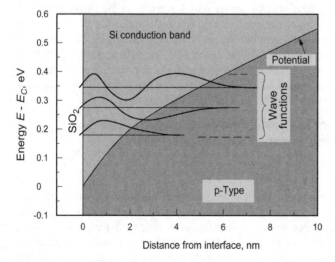

Fig. 2.11 Potential and wave functions in a MOS structure operated in inversion

$$\alpha = \frac{q^2}{2\varepsilon_0\, h\, c} \tag{2.2}$$

can be measured with great accuracy, as developed in [13] (ε_0 is the dielectric constant of vacuum, c the velocity of light).

2.5.2   Tunneling Currents

Other systems manufactured are sandwich structures (e.g., GaAlAs–GaAs–GaAl As). They are based on the fact that GaAlAs for instance, has a band gap of 2.0 eV while GaAs has a band gap of only 1.4 eV. By applying a voltage, the band structure is bent as schematically presented in Fig. 2.12. In a conventional consideration, no current is allowed to flow between the contacts ($x < 0$, $x > c$) irrespective of an applied voltage because the barriers ($0 < x < a$ and $b < x < c$) are to prevent this. However, by assuming considerably small values of magnitudes a, b, and c, a tunneling current can flow when the external voltage places the energy levels outside and inside (here E'') on the same value (in resonance). It should be noted that after exceeding this condition, the current sinks again (negative differential conductivity).

This sandwich structure is the basis for a good number of devices such as lasers, resonant tunneling devices or single-electron transistors. They will be treated in the section on electrical nanodevices.

2.6   Evaluation and Future Prospects

The state of the available molecule and cluster simulation programs can be described as follows: the construction of a molecule occurs under strict ab initio rules, i.e., no free parameters will be given which must later be fitted to experiment; instead, the Schrödinger equation is derived and solved for the determination of eigenenergies and eigenfunctions in a strictly deterministic way. The maximum manageable molecular size has some 100 constituent atoms. The limitation is essentially set by the calculation time and memory capacity (in order to prevent difficulties, semi-empirical approximations are also used. This is done at the expense of the accuracy). Results of these calculations are

- Molecular geometry (atomic distances, angles) in equilibrium,
- Electronic structure (energy levels, optical transitions),
- Binding energy, and
- Paramagnetism.

The deficits of this treatment are the prediction of numerous desired physical properties: temperature dependency of the above-named quantities, dielectric behavior, absorption, transmission and reflection in non-optical frequency ranges, electrical conductivity, thermal properties. However, there are attempts to acquire these properties with the help of molecular mechanics and dynamics [14–16].

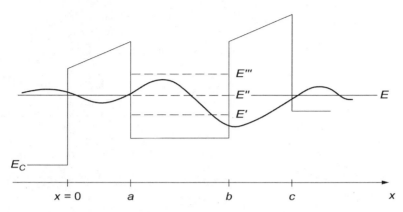

Fig. 2.12 Conduction band edge, wave function, and energy levels of a heterojunction by resonant tunneling

In numerous regards it is aim to bond foreign atoms to clusters. It is examined, for instance, whether clusters are able to bond a higher number of hydrogen atoms. The aim of this effort is energy storage. Another objective is the bonding of pharmaceutical materials to cluster carriers for medication depots in the human body. The above-named programs are also meant for this purpose. However, it should be stressed that great differences often occur between simulation and experiment, so that an examination of the calculations is always essential. Any calculation can only give hints about the direction in which the target development should run.

As far as the so-called quantum-mechanical influences on devices and their processes are concerned, the reader is kindly referred to the chapters in which they are treated. However, we anticipate that the investigation for instance, of current mechanisms in nano-MOS structures alone has given cause for speculations over five different partly new current limiting mechanisms [17]. The reduction of electronic devices to nanodimensions is associated with problems which are not yet known.

3 Nanodefects

3.1 Generation and Forms of Nanodefects in Crystals

The most familiar type of nanostructures is probably the nanodefect. It has been known for a long time and has been the object of numerous investigations. Some nanodefects are depicted in Fig. 3.1.

Their first representative is the vacancy, which simply means the absence of a lattice atom (e.g., silicon). In the case of a substitutional defect, the silicon atom is replaced by a foreign atom that is located on a lattice site. A foreign atom can also take any other site; then we are dealing with an interstitial defect.

It is a general tendency in nature that a combination of two or more defects is energetically more favorable than a configuration from the contributing isolated defects. This means that two (or more) vacancies have the tendency to form a double vacancy, triple vacancy etc., since the potential energy of a double vacancy is smaller than that of two single vacancies. The same reason applies to the formation of a vacancy/interstitial complex. It turns out also that a larger number of

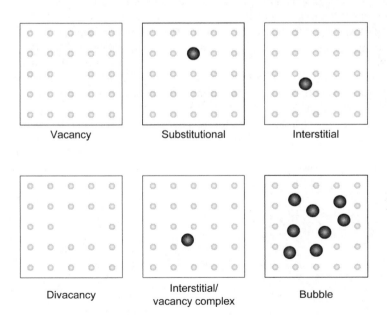

Fig. 3.1   Some nanodefects

vacancies can form a cavity in the crystal which can again be filled with foreign atoms so that filled bubbles are formed.

There is a long list of known defects; their investigation is worth the effort. Unfortunately, an in-depth discussion is beyond the scope of this book. Consequently, the interested reader is kindly referred to literature references such as [18] and [19] and quotations contained therein.

Defects in a crystal can result from natural growth or from external manipulation. At the beginning of the silicon age it was one of the greatest challenges to manufacture substrates which were free from dislocations or the so-called striations. Even today, semiconductor manufacturers spare no efforts in order to improve their materials. This particularly applies to "new" materials such as SiC, BN, GaN, and diamond. Nonetheless, research on Si, Ge, and GaAs continues. The production of defects takes place intentionally (in order to dope) or unintentionally during certain process steps such as diffusion, ion implantation, lithography, plasma treatment, irradiation, oxidation, etc. Annealing is often applied in order to reduce the number of (produced) defects.

3.2   Characterization of Nanodefects in Crystals

A rather large number of procedures was developed in order to determine the nature of defects and their densities. Other important parameters are charge state, magnetic moment, capture cross sections for electrons and holes, position in the energy bands, optical transitions, to name but a few. The following figures show some measuring procedures which explain the above-mentioned parameters.

Figure 3.2 describes an example of the decoration. This is based on the above-mentioned observation that the union of two defects is energetically more favorable than that of two separate defects. Therefore, if copper, which is a fast diffuser, is driven into silicon (this takes place via immersing the silicon into a $CuSO_4$ containing solution), it will be trapped by available defects. Cu is accessible by infrared measurements, while the available defects are invisible. The figure shows two closed dislocation loops and a third one inside shortly before completion. A dislocation can be explained by assuming a cut in a crystal so that n crystal

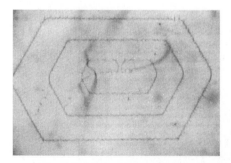

Fig. 3.2   Dislocations in Si doped with Cu [20]

planes end in the cut plane. $n + 1$ crystal planes may end on the other side of the cut. Then an internal level remains without continuation. Roughly speaking, the end line of this plane forms the dislocation, which can take the form of a loop.

We will now consider a case where silicon is exposed to a hydrogen plasma and subsequently annealed. The effects of such a treatment vary and will be discussed later. Here we will show the formation of the so-called platelet (Fig. 3.3).

A platelet is a two-dimensional case of a bubble, i.e., atoms from one or two lattice positions are removed and filled with hydrogen, so that a disk-like structure is formed (Fig. 3.4).

The proof of H_2 molecules and Si-H bonds shown in Fig. 3.4 can be done by means of Raman spectroscopy. This is an optical procedure during which the sample is irradiated with laser light. The energy of the laser quantum is increased or decreased by the interaction of quasi-free molecules with the incoming light. The modified reflected light is analyzed in terms of molecular energies which act as finger print of the material and its specific defects.

p-type Czochralski (Cz) Si is plasma-treated for 120 min at 250 °C and annealed in air for 10 min at temperatures between 250 °C and 600 °C. The Raman shift is measured in two spectral regions [22, 23]. At energies around 4150 cm^{-1} the response due to H_2 molecules is observed (Fig. 3.5a), and around 2100 cm^{-1} that due to Si-H bonds (Fig. 3.5b).

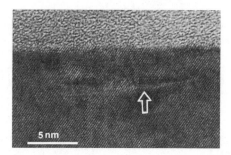

Fig. 3.3 Formation of a (100) platelet in Si by hydrogen plasma at 385 °C [21]. The image has been acquired by the transmission electron microscopy.

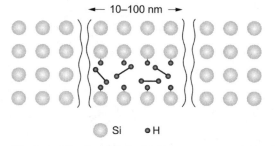

Fig. 3.4 Platelets filled with H_2 molecules and Si-H bonds (schematic)

It should be noted that after plasma exposure the surface is nanostructured and SiO_x complexes are formed there (Fig. 3.6). The p-type sample has been exposed to a hydrogen plasma for 120 min at 250 °C and annealed in air 10 min at 600 °C. The SiO_x complexes are detected with *photoluminescence*.

If oxygen-rich (e.g., Czochralski, Cz) material is exposed to a hydrogen plasma at approximately 450 °C, the so-called thermal donors are formed (most likely oxygen vacancy complexes). They can be measured with *infrared (IR) absorption*. The signal of the two types of thermal donors is shown in Fig. 3.7 [23].

Some defects possess magnetic moments (or spins) which are accessible by *electron spin resonance measurements*. Examples of systems which have been examined rather early are color centers in ionic crystals. Later, defects in GaAs have been of great interest. An example of the determination of the energy structure of the defects in GaAs is shown in Fig. 3.8 [24].

The MOS capacitance is an efficient tool for detecting defects in the oxide, in the neighboring silicon and at the Si/SiO_2 interface. We are limited to the discussion of defects in silicon, approximately in the neighborhood of 1 to 10 μm from the interface. If the (high frequency) capacitance is switched from inversion into deep depletion, it follows first the so-called pulse curve and then returns to the initial inversion capacitance at a fixed voltage (Fig. 3.9). It is generally assumed that the relaxation is controlled by the internal generation g within the depletion zone. It reflects a special case of the Shockley-Hall-Read generation recom-

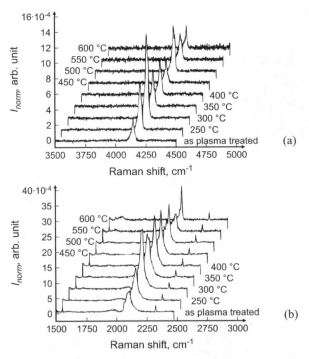

Fig. 3.5   Raman shift of H_2 bonds (a) and of Si-H bonds (b) [22]

bination statistics:

$$g = \frac{q\,n_i}{2\tau_G} \tag{3.1}$$

where n_i is the intrinsic charge carrier density and τ_G the generation lifetime

$$\frac{1}{\tau_G} = \frac{\sigma\,v_{th}\,N_T}{\cosh\dfrac{E_T - E_i}{k\,T}} \tag{3.2}$$

(σ: capture cross section, E_T: energy level, N_T: density of traps, E_i: Fermi level, and v_{th}: thermal velocity).

We can easily show that a plot of

$$-\frac{\mathrm{d}}{\mathrm{d}t}\left(\frac{C_{ox}}{C_{hf}}\right)^2$$

(the integrated generation rate G_{tot}, i.e., the generation current density) *vs.*

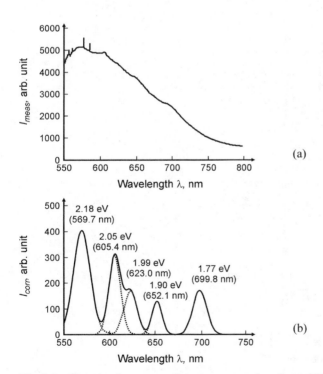

Fig. 3.6 Photoluminescence of a nanostructured surface of Si, (a) as measured, (b) after subtracting the background [23]

$$\frac{C_{hf,\infty}}{C_{hf}} - 1$$

(the normalized space charge depth W_g) from the data of Fig. 3.9b delivers a straight line. This plot is called Zerbst plot [25]. The slope of the straight line is

$$\frac{2C_{ox}\, n_i}{\tau_G\, N\, C_{hf,\infty}}$$

and thus inversely proportional to the generation lifetime. An example is given in Fig. 3.10.

However, the Zerbst plot is based on the fact that the lifetime is constant in the depth of the passage. If this is not the case, it is helpful to interpret the coordinates of the Zerbst plot anew so that the abscissa of the depth is the generating space-charge zone, $W_g = x = \varepsilon_{Si}/C_D$ (C_D is the depletion capacitance of silicon) while the ordinate is the generation current, i.e., the integral of the local generation rate over the momentary space charge depth, x. Then the differentiated curve delivers the local generation rate g, and like derived from the Shockley-Hall-Read statistics, a measure for the local density of the traps:

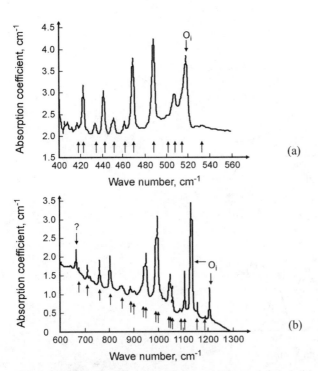

Fig. 3.7   IR absorption spectra for neutral thermal donors (a) and single-ionized thermal donors (b)

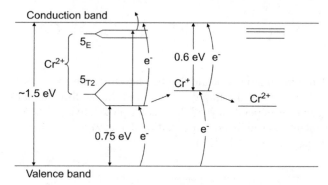

Fig. 3.8   Cr levels in GaAs

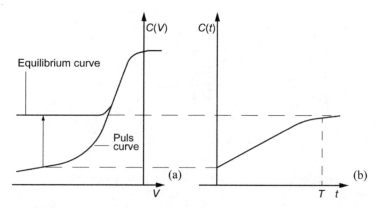

Fig. 3.9   MOS capacitance after switching from inversion in deep depletion (a) and during the relaxation (b)

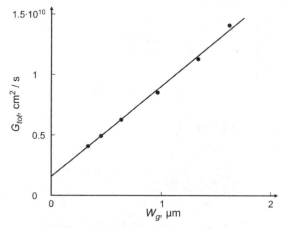

Fig. 3.10   Zerbst plot [26]

$$g(x) = \frac{q \, n_i \, v_{th} \, \sigma}{2 \cosh \dfrac{E_T - E_i}{k \, T}} \, N_T(x)$$ (3.3)

N_T and g are now considered as functions of the depth. Therefore, the differentiated Zerbst plot delivers a measure for the trap distribution in the depth while the measurement of the temperature dependency of the Zerbst plot delivers E_T. An example by which ion implantation induced traps (lattice damage) are measured and analyzed is given in Figs. 3.11 and 3.12 [27].

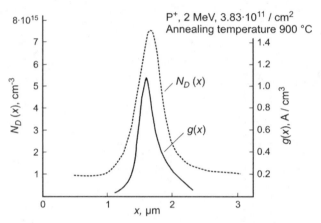

Fig. 3.11   Doping and generation rate profiles after phosphor implantation [27]

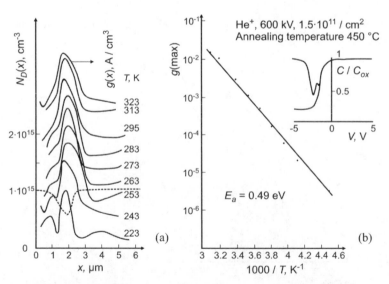

Fig. 3.12   (a) Generation rate profiles (full curves) and doping profile (dashed curve),
(b) after helium implantation and Arrhenius presentation of the generation rate [27]

Deep level transient spectroscopy (DLTS) is another helpful electrical procedure. It measures the trap densities, activation cross sections, and energy positions in the forbidden band. It is applied to Schottky and MOS diodes. The fundamentals are shown in Fig. 3.13.

The Schottky diode is switched from the forward to the reverse direction. Similarly, the MOS diode is switched from accumulation to depletion. After pulsing and retention of a fixed voltage it turns out that the capacitance runs back to a higher value. The summation of all pulse and relaxation capacitances produce the capacitance curves $C_\Pi(V)$ and $C_=(V)$. All information is obtained from the capacitance-transient $C(t)$, an exponential-like function.

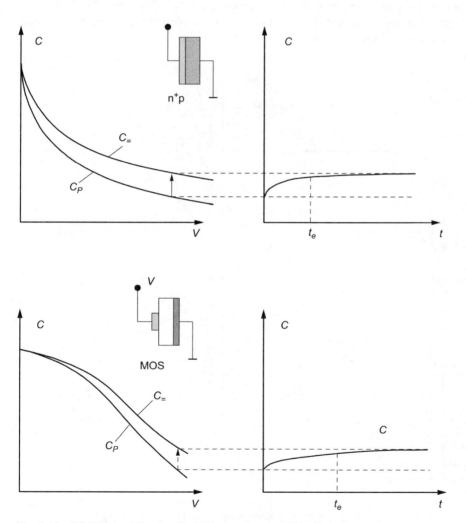

Fig. 3.13 DLTS on a Schottky diode (top) and on an MOS diode (bottom)

The reason for the capacitance relaxation is the presence of traps (for reasons of simplicity we will consider only bulk traps for the Schottky diode and surface states for the MOS capacitance). Figure 3.14 demonstrates the behavior of the traps after pulsing. The emission time constant τ_e is reflected in the capacitance relaxation of Fig. 3.13.

Technically, it is difficult to measure the full transient. In the worst case this would be called a speedy measurement of a maximum of a few femtofarad within a time span of less than one micro second. The measurement is done in such a way that two time marks are set for instance, at 1 and 2 ms. Then the transient is repeatedly measured within this window at different temperatures (Fig. 3.15).

It should be noted that in reality the capacitance $C(t \to \infty)$ is defined as zero and the negative deviation from $C(\infty)$ represents the signal. On the right of the figure, the capacitance difference $|C(t_1) - C(t_2)|$ is plotted against the temperature. The emission time constant (e.g., for electron emission) is

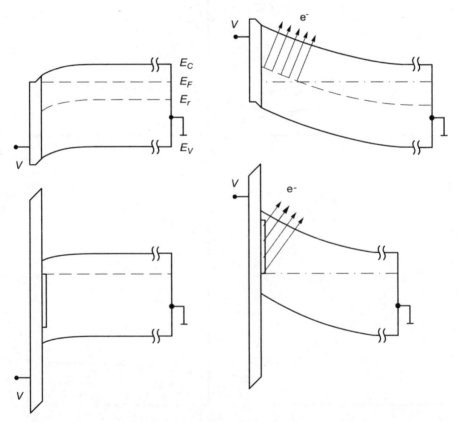

Fig. 3.14 Electron emission process after switching in the reverse state (Schottky, top) and depletion (MOS, bottom)

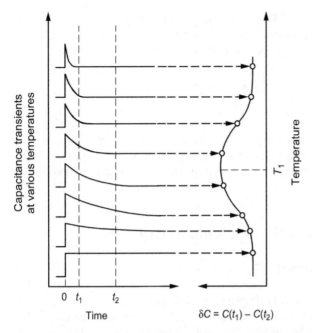

Fig. 3.15   Capacitance transients at various temperatures [26]

$$\tau_e = \frac{1}{c_n \, n_i} e^{\frac{E_T - E_i}{kT}} = \frac{1}{v_{th} \, \sigma_n \, N_C} e^{\frac{E_T - E_C}{kT}}$$ (3.4)

or

$$T^2 \tau_e = \frac{1}{\gamma_n \, \sigma_n} e^{\frac{E_T - E_C}{kT}}$$ (3.5)

(c_n is the capture constant of the emitting traps, σ_n the capture cross section, n_i the intrinsic density, E_T the position of the traps in the band gap, E_i the intrinsic Fermi level, and v_{th} the thermal velocity). At very low temperatures the emission time is high compared to the time window $t_1 - t_2$. At very high temperatures the reverse applies, so that the transient is finished long before t_1 is reached. In between, there is a maximum δC_{max}, at the temperature T_{max}. For a given window t_1, t_2, the emission time at this maximum is calculated as

$$\tau_e = \frac{t_2 - t_1}{\ln(t_2 / t_1)}$$ (3.6)

Now a data pair ($T_{max}^2 \, \tau_e$, T_{max}) is available and can be substituted in Eq. 3.1. The same procedure is repeated for other time windows, so that a curve $T_{max}^2 \, \tau_e$ vs. T_{max} and thus, the energy $E_C - E_T$, i.e., the position of the trap energy in the forbidden band can be determined. From the same equation, the unknown quantity σ_n can be determined. In order to describe the determination of the trap density, we will use the example of the Schottky diode. It is shown that N_T is given by

$$N_T = 2N_D \frac{\delta C_{max}}{C_0} \frac{(t_2/t_1)^{\frac{t_2}{t_1}/\left(\frac{t_2}{t_1}-1\right)}}{1-t_2/t_1} \tag{3.7}$$

An example of the measurement of a capacitance transient is given in Fig. 3.16, where the emission is detected from two traps [28]. A second example of the analysis of the activation energy of the two traps of Fig. 3.16 is presented in Fig. 3.17 [28].

3.3   Applications of Nanodefects in Crystals

3.3.1   Lifetime Adjustment

An essential parameter of a power device (e.g., a thyristor) is the time required to switch it from the forward to the reverse state. This time is measured by re-switching the voltage over the device from the forward to the reverse state (Fig. 3.18).

The storage time t_s is determined mainly by recombination in the base and thus by the carrier lifetimes, τ_n and τ_p. As a rule of thumb, t_s can be expressed by means of the equation

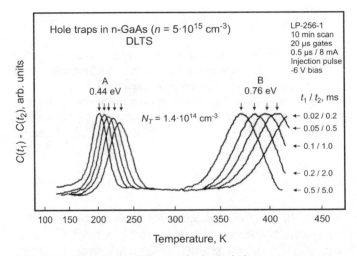

Fig. 3.16   Capacitance difference in time window *vs.* temperature

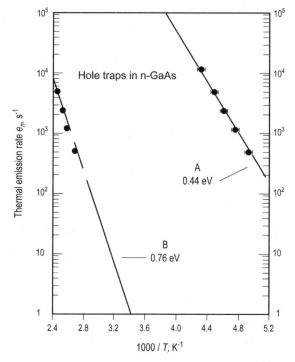

Fig. 3.17 Activation energies of the two traps in Fig. 3.16 derived from Eqs. 3.5 and 3.6

$$\mathrm{erf}\sqrt{\frac{t_s}{\tau_r}} = \frac{1}{1 + I_R / I_F} \qquad (3.8)$$

τ_r is identical to one of the minority carrier lifetimes, τ_n and τ_p, or to a combination of both. Therefore, each attempt to accelerate the switching times must concentrate on the shortening of the lifetimes τ_n and τ_p. Technically this is achieved by the introduction of point defects or defects of small dimensions in the critical zone of the semiconductor. N_T is assumed as their density and c_n or c_p their probability of capture of electrons or holes (c_n and c_p are related to the initial cross section $c_{n,p} = v_{th}\,\sigma_{n,p}$, where v_{th} is the thermal velocity). The theory of Shockley, Hall, and Read shows that the lifetime is related to the number of traps by

$$\tau_{n,p} = \frac{1}{c_{n,p}\,N_T} \qquad (3.9)$$

Traps in a power device also lead to unfavorable effects. This includes the rise of the forward resistance and the leakage current in the reverse state.

The traps can be brought into the semiconductor in different ways. Early procedures have been gold or platinum doping, or electron and gamma ray exposure. Today, best results are obtained by hydrogen implantation.

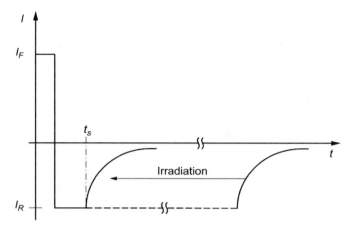

Fig. 3.18 Reverse recovery of a thyristor. Virgin device (right curve), and after lifetime shortening (left curve)

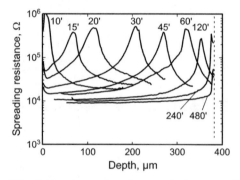

Fig. 3.19 Formation of a p-n junction through thermal donors [30]. *2-step process*

3.3.2   Formation of Thermal Donors

Since the sixties it is well-known that heating of (oxygen rich) Czochralski (Cz) silicon leads to the transformation of oxygen into a defect, which acts as donor in silicon. Normally that is a slow process and the donors disappear at temperatures above 450 °C. Therefore, these donors are of no technical interest. However, recently they have attracted new attention because their production process can be accelerated by some orders of magnitude: the Cz wafer is exposed to a hydrogen plasma before annealing. An example is shown in Fig. 3.19. In this example, the p-type wafer is subjected to a 110 MHz plasma of 0.35 W/cm² at a hydrogen pressure of 0.333 mbar and a temperature of 250 °C. Then the wafer is annealed at 450 °C in air for the various times noted in the figure. The wafer is beveled and the spreading resistance measured (in addition also see Sect. 4.2.4). The maxima display the positions of the p-n junctions which are formed from the original p-type material (on the right of the maximum) and the newly formed n-type material

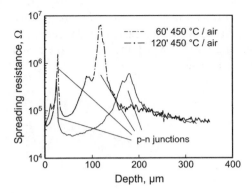

Fig. 3.20   Formation of a three-layered structure in exposed silicon [31]

(left). Very high penetration rates are achieved: after four hours the 380 μm thick wafer is completely converted into n-type. A conversion depth of 80 μm for instance, is received after 16 min. The same depth is obtained for a classical diffusion after 16 h (Ga, 1300 °C).

If the penetration depth *vs.* the annealing time is plotted, the diffusion constant $D = 2.9 \cdot 10^{-6}$ cm^2/s can be derived. This number is comparable with the prediction of van Wieringen and Warmholz [29] for the diffusion constant of atomic hydrogen.

The limited thermal budget of these p-n junctions and the devices made from them can be clearly extended if a procedure is used which leads to the so-called *new* thermal donors.

Another variant is obtained if denuded silicon is used. This is the fundamental material for electronic devices: a thin layer is denuded from oxygen (by simple outdiffusion), while in two subsequent steps of nucleation and precipitation, the inside of the silicon is prepared as gettering zone for impurities. Thus a high-grade cleaned surface zone remains for the manufacturing of electronic circuits.

After the application of hydrogen plasma and the subsequent annealing for the formation of thermal donors in this material, a double peak appears upon measurement of the spreading resistance (Fig. 3.20). As an explanation, it should be remembered that the transformation to n-type requires the use of both oxygen (the later thermal donors) and hydrogen (as catalyst). On the right of the second maximum, a p-type behavior of the raw material is observed. The hydrogen could not reach this region. The zone between the two maxima is converted; oxygen and hydrogen are available. Before the first maximum no oxygen is available for the transformation in the exposed zone. Some types of devices can be expected on the basis of this structure.

3.3.3   Smart and Soft Cut

One of the fundamental problems of the production of integrated circuits lies in the mutual isolation of passive and active devices which are built on the surface of the semiconductor. Usually the problem is solved in such a way that a first isolation separates all surface circuits from the bulk while in a second step electrically isolated islands are formed on the remaining layer in which the individual circuits are contained. In the following we treat the first step.

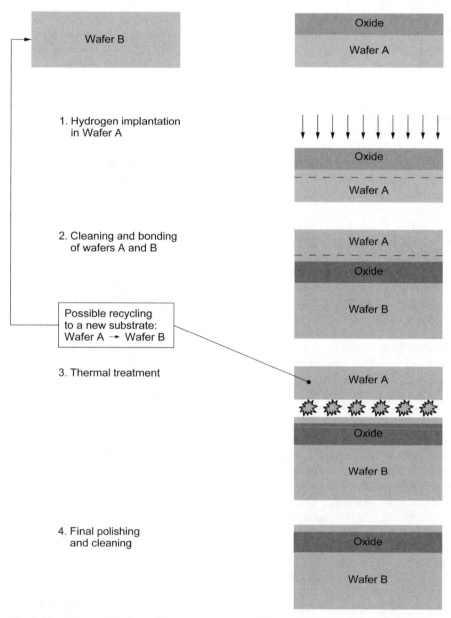

Fig. 3.21   Smart cut (schematic)

Early solutions to bulk isolation have been epitaxial deposition of the active layer, e.g., n-type on p-type substrate, silicon-on-sapphire (SOS), and silicon-on-oxide (SOI). The latter includes versions like (i) oxygen implantation and SiO_2 formation, (ii) deposition of amorphous Si on SiO_2 and recrystallization, and (iii) wafer bonding. All show specific pro and cons.

A newly established SOI procedure is based on the formation of point defects (Fig. 3.21). A first wafer, A, is oxidized and implanted with hydrogen through the oxide.

The implantation energy is selected in such a way that the ions come to rest under the SiO_2/Si interface after a few micrometers. The wafer is now placed head-first on a second wafer, B, so that the oxide comes in close contact with wafer B. By suitable annealing, this contact is intensified and simultaneously the wafer A splits at the place where the ions come to rest. After removing the main part from wafer A, a configuration of silicon (wafer B), oxide, silicon-on-oxide (a remaining thin layer of wafer A) remains. The active circuits are then manufactured on the thin layer. The applied doses are about 10^{17} cm^{-2}. It is shown [32] that plasma hydrogenation reduces the required dosed by a factor of 10 (Fig. 3.22). This procedure is called soft cut.

After hydrogen implantation the wafer is subsequently annealed (1000 °C, H_2 atmosphere) and, more importantly, hydrogenated with plasma. It is evident that a dose of a few 10^{16} cm^{-2} and the hydrogenation to a maximum concentration produce the required 10^{21} H/cm^{-3}. A saving within an order of magnitude is an enormous gain in the production costs since the implantation is much more expensive than the hydrogenation.

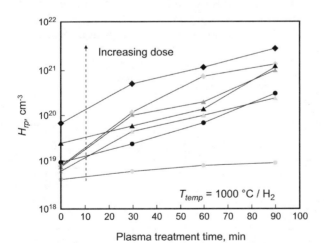

Fig. 3.22   Increase of hydrogen in the maximum position of the implantation profile by hydrogenation. Black symbols: $1\cdot10^{15}$, $1\cdot10^{16}$, and $3\cdot10^{16}$ cm^{-2} hydrogen dose, $E = 70$ keV, dark-gray symbols: $1\cdot10^{16}$ cm^{-2} helium dose, $E = 300$ keV, light-gray symbols: $1\cdot10^{15}$, $1\cdot10^{16}$, and $1\cdot10^{17}$ cm^{-2} helium dose, $E = 1$ MeV. The concentrations were determined by secondary ion mass spectroscopy [32].

3.3.4   Light-emitting Diodes

(i) *Nanoclusters in SiO₂.* It has been a long-nourished hope to develop opto-electronic components by means of standard silicon technology. However, silicon is not suitable because of its indirect band gap. Therefore, the electroluminescence discovered on SiO_2 by implantation with Ge or Si is examined with great effort. Low-temperature (120–150 °C) and dose values are selected for the implantation in such a way that an average surplus density Si or Ge is set to a few atomic per cent. The wafers are then annealed in N_2 at 1000 °C for 30 to 60 min. Strong photoluminescence (PL) and electroluminescence (EL) spectra are observed (Fig. 3.23). EL is caused by a current of 100 nA / mm² at an applied voltage of 370 V [33].

Although complex models have been developed to explain the phenomenon [34], many details still remain unclear. However, it is generally assumed that the mechanism is based on a quantum confinement effect of reconstructed nanoclusters.

The EL capability is used for the building of an optoelectronic coupler which contains the light-emitting device from the above-mentioned implanted oxide (Fig. 3.24). The detector is based on amorphous silicon. Both sections can be produced with standard silicon technology.

(ii) *Porous silicon.* Chemical and electrochemical etching [36 and literature quoted therein] and ion implantation [37] are used for the production of porous (po-Si) or porous-like silicon. Then a typical LED structure can be formed by making a p-type wafer porous on a surface which is covered by indium–tin–oxide (ITO) and a metal electrode in form of a finger. The back side is fully metallized in order to acquire an ohmic contact. A positive voltage may be applied at the metallized front. There are numerous versions of porous LED structures including those from homo-pin or epitaxial heterojunction structures. A typical EL spectrum can be seen in Fig. 3.25.

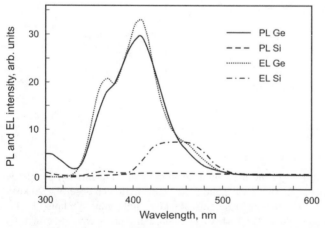

Fig. 3.23   Electroluminescence of an implanted MOS oxide [33]

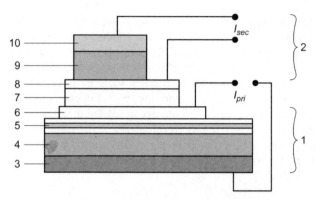

Fig. 3.24 EL from a nanocluster based optoelectronic coupler [35]. 1: LED, 2: detector, 3 and 10: metallic contacts, 7: optical transparent galvanic isolation layer, 4: Si wafer, 5: SiO_2 layer with implanted nanoclusters, 6: optical transparent conductive layer, wafer back contact, 8: optical transparent conductive layer, 9: pin a-Si photodiode

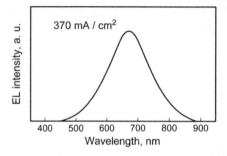

Fig. 3.25 Electroluminescence from porous silicon [38]

The technology of porous silicon has always been connected with the hope of combining PL with standard silicon technology. These hopes were soon subdued by the low conversion coefficients, some of which are in the 10^{-5} range. However, the full limits of the applicability of po-Si are not yet known.

(iii) It is well-known that the interaction of silicon, oxygen, and hydrogen leads to EL and PL within the optical range. Section 3.2 shows a CZ wafer that is exposed to a 13.56 MHz hydrogen plasma at 250 °C for 2 h, followed by a 10 min oxidation at 600 °C (i.e., exposure to air). Thereafter the wafer shows a strong EL in addition to the already available PL shown in Fig. 3.6.

3.4 Nuclear Track Nanodefects

3.4.1 Production of Nanodefects with Nuclear Tracks

In the process of irradiating insulators with high-energy ions (typical energies of 100 MeV to 1 GeV) a change of the material in the path area was observed. The density of the material decreases in a cylinder by a few nanometers around the ion trajectory while the density increases at the edge of the cylinder. Concomitantly, different characteristics also change; thus, diamond for instance, which otherwise does not accept foreign atoms by diffusion can be doped at these positions.

This concept has been applied in particular to plastics such as polyethylene-terephthatalate (PET), polydimethylsiloxane, polyaniline, polyethylenedioxythiopene. Foils of the material with a thickness of 10 μm are irradiated in a heavy-ion accelerator. The so-called latent nuclear tracks develop. For further application these are opened by etching. The etching procedures vary from material to material; e.g., a 5 molar NaOH etch is well suitable for PET at 60 °C. The result of this treatment is shown in Fig. 3.26 [39].

3.4.2   Applications of Nuclear Tracks for Nanodevices

The possibility to process etched nuclear tracks in the nanometer range is their fundamental attraction.

A probable application would be the metallic sealing of the developed hole at the front and back sides and filling the cavity with a gas. Thus, nanometric plasma displays could be manufactured (Fig. 3.27).

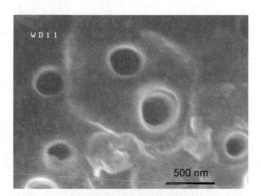

Fig. 3.26   Nuclear track in a PET foil. Layer thickness 10 μm; irradiated with 2.5 MeV/u
^{84}Kr (i.e., 210 MeV acceleration energy)

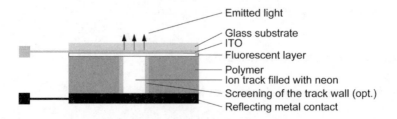

Fig. 3.27   Nanometric light emitting rod

Another concept is the sequential coverage of the inner cylinder walls with metals and insulators. This is a way to produce cylinder capacitors.

By using the nuclear tracks as a via, coils and inductances can be manufactured, provided that they are arranged in a skillful way. By combining capacitors, coils and hybrid applied plastic electronics are even conceivable as complete analog circuits.

3.5   Evaluation and Future Prospects

Ever it has been possible to grow suitable semiconductor materials for electronic devices, the defects contained in the material have been regarded as hostile and harmful. On the whole this finding is still correct but in the meantime, niches have developed in which defects deliver positive applications. The first example is of course the procedure described above for the switching time adjustment of power devices. Although it has been worked on for more than 30 years, it is still the subject of intensive investigations [40]. Historically, the next application is the back side gettering which works with different methods such as back side implantation, mechanical graining, coverage with phosphorus silicates etc. [41]. The idea common to all procedures is that the defects of the back side are supposed to attract impurities inside the silicon and catch them permanently. Today's solution is based on the same principle, even if the getter center is now inside the silicon. Moreover, this procedure is still investigated thoroughly despite certain experiences by manufacturers of semiconductor material. With procedures such as smart and soft cut, nanodefects play a new role in the device production. They are directly used for the production of certain structures. In process engineering, this procedure is referred to as *defect engineering.* In the meantime, smart cut has found a parallel application in solar cell production [42]: the surface of originally monocrystalline silicon is converted into porous silicon by current. This occurs by forming two thin layers of different properties. In particular the upper layer can be recrystallized by for instance, a laser treatment while the lower one remains porous. This lower layer is removable from the wafer so that a thin layer structure is gained which can be applied on a ceramic substrate for further treatment. In this way the economical production of many thin layer solar cells from one wafer is desirable. In the whole area of photovoltaics, defects which are produced by the exposure of silicon in a hydrogen plasma are expected to substantially improve the properties of the solar cell. This applies particularly to the surface whose free silicon bonds are to be saturated by hydrogen.

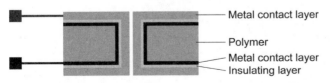

Fig. 3.28 Nanometric capacitor

In diamond electronics, hydrogen is used in order to obtain p-type conductivity [43]. Hydrogen contributes to the improvement of the results achieved so far. Moreover, the above-mentioned surface stabilization is used by way of trial on diamond [44].

The following developments are in progress in the area of nuclear tracks:

- Production of new electrodes for electrochemistry [45]
- Incorporable medication containers for long-term supply [45]
- Quantum diodes [46]
- Thermoresponsive valves [47]
- Nanometric light-emitting diodes [48]
- Micro-inductances for oscillators in communication technology [49]
- Micro-photodiodes [50]
- Pressure and vapor sensors [51]
- Transistors [52]

4 Nanolayers

4.1 Production of Nanolayers

4.1.1 Physical Vapor Deposition (PVD)

In general, physical vapor deposition (PVD) from the gas phase is subdivided into four groups, namely (i) evaporation, (ii) sputtering, (iii) ion plating, and (iv) laser ablation. The first three methods occur at low pressures. A rough overview is seen in Fig. 4.1.

(i) *Evaporation.* This procedure is carried out in a bell jar as depicted in Fig. 4.2. A crucible is heated up by a resistance or an electron gun until a sufficient

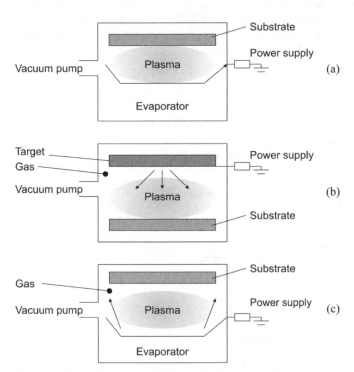

Fig. 4.1 Three fundamental PVD methods: evaporation (a), sputtering (b), and ion plating (c) [53]

vapor pressure develops. As a result, material is deposited on the substrate. Technically, the resistance is wrapped around the crucible, or a metal wire is heated up by a current and vaporized. The electron gun *(e-gun)* produces an electron beam of, e.g., 10 keV. This beam is directed at the material intended for the deposition on the substrate. The gun's advantage is its unlimited supply of evaporating material and applicability of non-conductive or high-melting materials. Its shortcomings lie in the production of radiation defects, for instance in the underlying oxide coating. Both procedures are more precisely depicted in Fig. 4.3.

(ii) *Sputtering.* In literature, there is no clear definition of the term *sputtering.* Generally, an atom or a molecule, usually in its ionized form, hits a solid state *(target)* and knocks out surface atoms. This erosion is accompanied by a second process, namely the deposition of the knocked out atoms on a second solid state *(substrate).* The latter process is relevant when forming thin layers.

(a) Glow discharge. In its simplest form, sputtering is achieved by glow discharge with dc voltage. A cross section of the arrangement is schematically represented in Fig. 4.4. After mounting the samples on a holder, the chamber is rinsed repeatedly with Ar. Eventually, a constant gas pressure of some 100 mPa is built up. The target, being attached a few centimeters above the substrate, is raised to a negative dc potential from −500 to −5 000 V, while both chamber and substrate are grounded. The discharge current requires a conducting target.

When the voltage is slowly increased, a small current flows over the two electrodes. This current is caused by the ions and electrons which normally appear in the gas and by the electrons which leave the target after ion bombardment (secon-

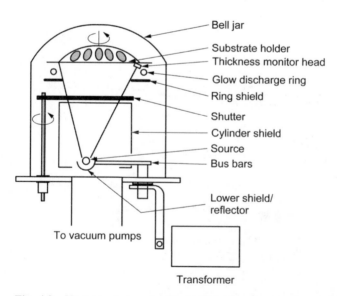

Fig. 4.2 Vacuum system for the vaporization from resistance-heated sources. When replacing the transformer and heater with an electron gun, vaporization by means of an electron beam occurs [54].

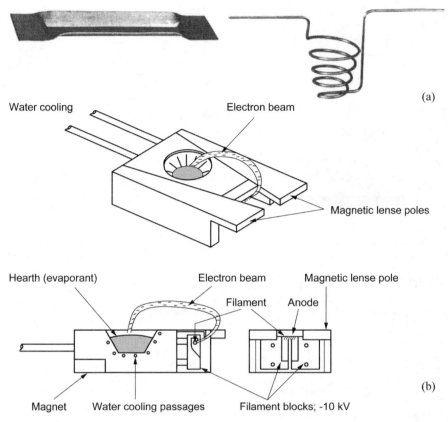

Fig. 4.3 Evaporation by means of resistance-heating with a tungsten boat and winding (a) and electron gun (b) [55]

dary electrons). At a certain voltage value, these contributions rise drastically. The final current-voltage curve is shown in Fig. 4.5.

The first plateau (at 600 V in our example) of the discharge current is referred to as Townsend discharge. Later the plasma passes through the "normal" and "abnormal" ranges. The latter is the operating state of sputtering. A self-contained gas discharge requires the production of sufficient secondary electrons by the impact of the ions on the target surface and conversely, the production of sufficient ions in the plasma by the secondary electrons.

(b) High frequency discharge. When replacing the dc voltage source from Fig. 4.4 with a high frequency generator (radio frequency, RF, generator), target and substrates erode alternately depending on the respective polarity. But even with these low frequencies, a serious shortcoming becomes apparent: due to the substantially small target surface (compared to the backplate electrode consisting of the bell, the cable shield, etc.) a proportionally large ion current flows if this backplate electrode is negatively polarized. This would mean that the substrates are covered with the material of the bell, which is not intended.

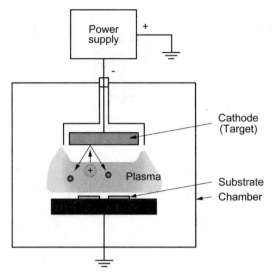

Fig. 4.4   DC voltage sputtering [56]

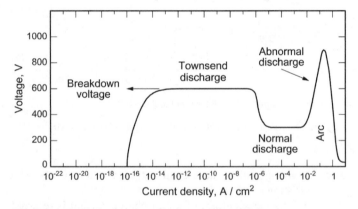

Fig. 4.5   Applied voltage *vs.* discharge current [56]

In order to overcome this shortcoming, a capacitor is connected in series be-tween the high frequency generator and the target, and/or the conducting target is replaced with an insulating one. During the positive voltage phase of the RF sig-nal, the electrons from the discharge space are attracted to the target. They impact on the target and charge it; current flow to the RF generator is prevented by the capacitor. During the negative half-wave of the RF signal, the electrons cannot leave the target due to the work function of the target material. Thus, the electron charge on the target remains constant.

Due to their mass, the positively charged ions are not capable of following the RF signal with frequencies above 50 kHz. Therefore, the ions are only subjected

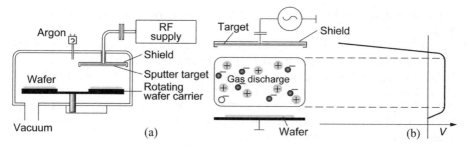

Fig. 4.6 (a) RF sputter system and (b) distribution of the potential in an RF plasma

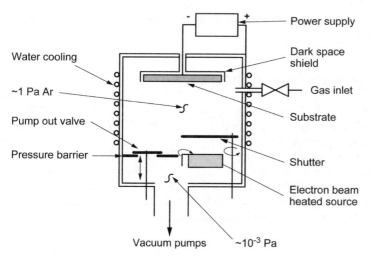

Fig. 4.7 Ion plating system [57], slightly modified

to the average electrical field which is caused by the electron charge accumulated on the target. Depending on the RF power at the target, the captured charge leads to a bias of 1 000 V or more and causes an ion energy within the range of 1 keV.

When using a capacitively coupled target, the limitations of the glow discharge can be overcome, i.e., a conducting target is no longer required. Therefore, the number of layers which can be deposited by sputtering is greatly increased.

(iii) *Ion plating*. This process is classed between resistance evaporation and glow discharge. A negative voltage is applied to the substrate, while the anode is connected with the source of the metal vaporization. The chamber is subsequently filled with Ar with a pressure of a few Pa, and the plasma is ignited. After cleaning the wafer by sputtering, the e-gun is switched on and the material is vaporized. The growing of the layer on the substrate is improved by the plasma in some properties such as adhesion and homogeneity compared to a sole PVD.

The advantages of ion plating are higher energies of the vaporized atoms and therefore better adhesion of the produced films. The disadvantage is heating of the substrate and plasma interactions with radiation-sensitive layers such as MOS oxides.

(iv) *Laser ablation*. The following process data are typical values. A high-energy focused laser beam (100 mJ, 1 J/cm^2) is capable of eroding the surface of a target rotating with a velocity of one revolution per second. The material is vaporized on the substrate, and as a result, a film is produced on it at a rate of 0.07 nm/laser pulse. The growth can be supported by heating the substrate (750 °C) and by chemical reactions (oxygen at 50 Pa). So far, the used lasers are excimer, Nd:YAG, ruby, and CO$_2$ lasers.

Advantages of laser ablation are the deposition of materials of high-melting points, a good control over impurities, the possibility of the vaporization in oxidizing environments, and stoichiometric vaporization. A shortcoming is the formation of droplets on the vaporized layer. A system described in the literature is presented in Fig. 4.8.

4.1.2   Chemical Vapor Deposition (CVD)

The CVD process is performed in an evacuated chamber. The wafer is put on a carrier and heated to a temperature between 350 and 800 °C. Four possible versions of the chamber are presented in Fig. 4.9.

One or several species of gases are let in so that a gas pressure is formed between very low and normal pressure. The gas flow hits the wafer at a normal or a glancing incidence. Now a dissociation (in the case of a single gas species) or a reaction between two species takes place. In both cases, a newly formed molecule adheres to the wafer surface and participates in the formation of a new layer. Let

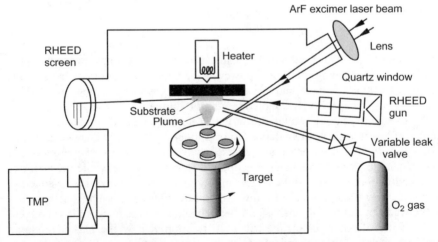

Fig. 4.8 Typical laser ablation system under O$_2$ partial pressure [58]. Note the so-called plume, a luminous cloud close to the irradiated target surface. RHEED: reflection high-energy electron diffraction

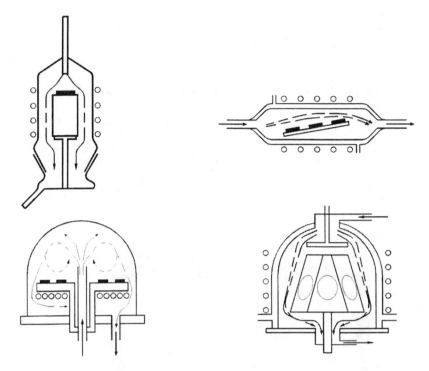

Fig. 4.9   Four versions of a CVD chamber

us consider silane (SiH_4) as an example of the first case. On impact, it disintegrates into elementary silicon, which partly adheres to the surface, and to hydrogen, which is removed by the pumps. The second case is represented by SiH_4, which reacts with N_2O to form SiO_2. The process can of course be accompanied by other types of gases which act as impurities in the deposited layer. Examples are phosphine (PH_3) or diborane (B_2H_6), which also disintegrate and deliver effective phosphorus or boron doping of the deposited silicon.

In this book, the closer definition of CVD, i.e., a layer structure without the continuation of the underlying lattice, is used. The reverse case is called vapor phase epitaxy. In some publications, both expressions are used without any distinction.

CVD deposition can be supported by an RF plasma, as schematically shown in Fig. 4.10, an example of an amorphous or micro-crystalline silicon deposition. The major difference to the conventional CVD is the addition of Ar for the ignition of the plasma and of H_2. The degree of the SiH_4 content in H_2 determines whether amorphous or microcrystalline silicon is deposited. In the first step, both types are deposited. However, a high concentration of H_2 etches the amorphous portion, and only the microcrystalline component remains. The etching process is even more favored if higher frequencies (e.g., 110 MHz) other than the usual 13.56 MHz are used. In Fig. 4.11, a typical PECVD system is depicted.

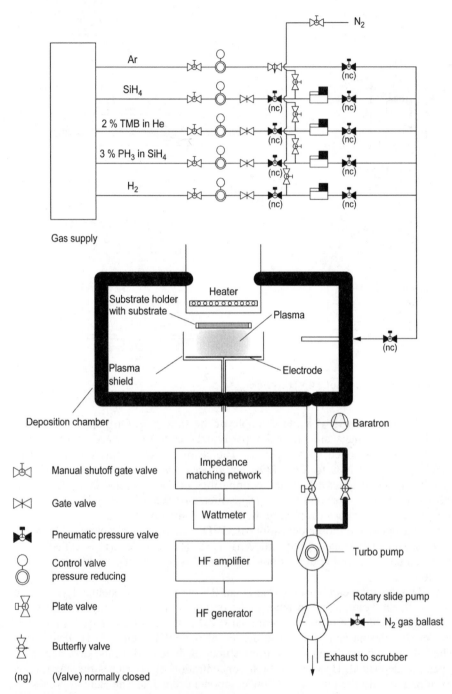

Fig. 4.10   Block diagram of a PECVD system

4.1.3   Epitaxy

We are dealing with epitaxy if a layer is deposited on a (crystalline) substrate in such a way that the layer is also monocrystalline. The layer is often referred to as film. In many cases, the film takes 99.9 % of the entire solid state, as in the example of a Czochralski crystal, which is pulled from a narrow seed nucleus. If film and substrate are from the same material, we are dealing with homoepitaxy (e.g., silicon-on-silicon), otherwise with heteroepitaxy (e.g., silicon-on-sapphire). Another distinction is made by the phase from which the film is made: vapor phase epitaxy, liquid phase epitaxy (LPE), and solid state epitaxy. A subclass of vapor phase epitaxy is molecular beam epitaxy (MBE).

The setup of a vapor phase epitaxy is not shown because it resembles the CVD setup shown previously to a good extent. However, the setup of a molecular beam epitaxy (MBE) is depicted in detail in Fig. 4.12.

Fig. 4.11   PECVD system for the deposition of amorphous solar cells

The constituents of the deposited film are contained in mini furnaces as elements, the so-called Knudsen cells, which are discussed below. During heating some vapor pressure develops and an atom beam is emitted, which is bundled by successive apertures. The beam hits the wafer surface to which the atoms remain partially adhered. There, they can react with atoms of a second or third beam, which is also directed towards the wafer surface. A favorable reaction and finally the film deposition depend on the selection of the parameters, i.e., wafer temperature, the ratios of the beam densities, the purity of the surface, etc. As shown in the same figure, the effusion cell can be replaced by an evaporation with the electron gun. The chamber contains many devices for the in situ inspection of the growing layers, for example low energy electron diffraction (LEED), secondary ion mass spectroscopy (SIMS), and Auger and Raman spectroscopy. The quality of the vacuum is controlled by a residual gas analyzer. The effusion (Knudsen) cell is seen in detail in Fig. 4.13.

The material to be deposited is contained in the innermost cell which is heated up. Its temperature is controlled by a set of thermocouples and resistance heaters. Without further measures, the high temperature leads to molecular desorption from all warmed up surfaces, to the emission of impurities into the substrate, and in the worst case, to the breakdown of the vacuum. Therefore, a screen cooled with liquid air is installed around the internal cell. Conversely, in order to avoid high thermal flows between furnace and screen, a water-cooled shield is inserted between them.

For the operation of the LPE, knowledge of the phase diagram is required. Let us consider the phase diagram of Ga and As as an example (Fig. 4.14). This phase

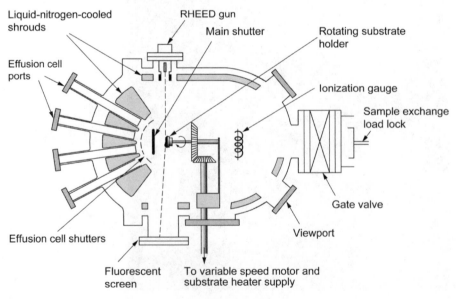

Fig. 4.12 Schematic structure of the MBE as in [59]

Thermocouples

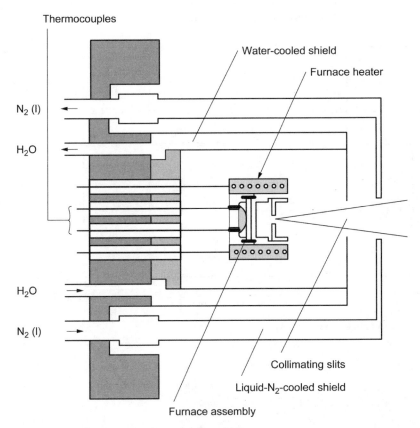

Fig. 4.13   Effusion cell (schematic) from [60]

diagram represents the case of a congruent phase transition. The mixture of Ga and As in a ratio of 50:50 takes over the role of a constituent, i.e., the phase diagram disintegrates into a first Ga-GaAs and a second GaAs-As phase diagram. In other words: GaAs forms a stable compound which is not subjected to any chemical change during the change of temperature.

Obviously, both diagrams are eutectic. Let us begin with a melt of Ga_xAs_{1-x} and a solid state of GaAs, both in close contact. In order to avoid sublimation of the As, the process should begin with a melt enriched with Ga so that we work on the left side of the phase diagram above the liquidus. Pure GaAs freezes out on the already available solid GaAs by decreasing the temperature (this is also the aim of LPE) and the melt becomes richer in Ga until finally only pure Ga remains. In the reverse case (operation on the right of GaAs), solid GaAs is again deposited but the melt is rich in As. Doping atoms can be added to the melt before the deposition, for instance, when fabricating p-n junctions. As an example, a GaAs wafer (the original solid state in the above example) is depicted in Fig. 4.15. The wafer is laterally pulled under different melts, so that the above mentioned deposition

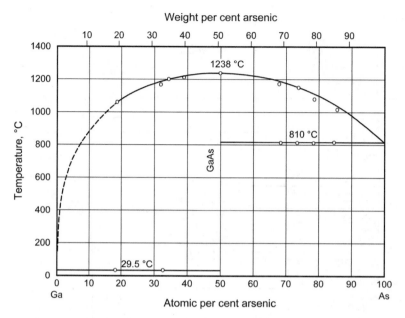

Fig. 4.14   Ga-As phase diagram [61]

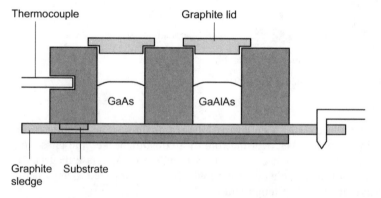

Fig. 4.15   Sledge arrangement for the fabrication of injection lasers using LPE

process can be repeated with different material layers (Ga, GaAlAs) and different doping types. This process was the early way of fabricating an injection laser.

Solid state epitaxy is a phenomenon which appears, for example, as a recrystallization of amorphous silicon. As is well known, ion implantation destroys the crystalline structure of silicon up to complete amorphization. If the wafer is annealed, the structure can be reorganized beginning from the unimpaired bulk of the wafer. This phenomenon is of less interest for nanotechnology.

It should be mentioned that there are some versions of the classical epitaxial procedures as *in-situ* implantation or laser-enhanced deposition.

Finally, let us compare some typical parameters of the different types of epitaxy (Table 4.1). The best control of the thickness, highest purity, and largest doping gradients are obviously obtained by MBE. These advantages are compensated by high equipment costs and low growth rates.

Table 4.1 Comparison of epitaxial parameters [60]

Process characteristic	LPE	VPE	MBE
Possibility of in situ etching	Yes, through melt back	Yes, through halide reaction with substrates above growth temperatures	No
Further cleaning and monitoring of substrate surface	Not possible	Possible by heat treatment in inert gas, but surface cannot be monitored except by ellipsometry	Yes, by ion bombardment or thermally in UHV. Can be monitored by AES, LEED or RHEED, but there may be electron beam effects
Typical growth rate range	0.1–1.0 μm/min	0.05–0.3 μm/min	0.001–0.03 μm/min
Layer thickness control	±50 nm	±25 nm	Easily ±5 nm, can be ±0.5 nm
Substrate temperature (for growth of GaAs on GaAs)	1120 K	1020 K	820 K
Interface control	Segregation and outdiffusion can occur	Autodoping and outdiffusion can occur	Only outdiffusion, but this may occur at enhanced rates under some conditions
Topography	Very difficult to obtain uniformly smooth surfaces over large areas	Can be very smooth but conditions for success are somewhat critical	Extremely smooth surfaces obtained under not very critical conditions. Even initial surface roughness is smoothened out.
Composition control of ternaries and quaternaries	Composition determined by process chemistry	Composition determined by process chemistry	Group III element ratio determined by thermal stability of the source. Group IV ratio by surface chemistry
Total carrier concentration in undoped film	Very low, $(N_D + N_A) \approx 10^{13}$ cm^{-3}	Low, $(N_D + N_A) \approx 10^{14}$ cm^{-3}	Rather high, $(N_D + N_A) \geq 10^{16}$ cm^{-3}

Table 4.1 (cont'd)   Comparison of epitaxial parameters [60]

Process characteristic	LPE	VPE	MBE
Presence of deep levels	Yes, $N_T \approx 10^{12}–10^{14}$ cm^{-3}, low end of range fairly easily realizable	Yes, $N_T \approx 10^{12}–10^{14}$ cm^{-3}, low end of range fairly easily realizable	Yes, $N_T \approx 10^{12}–10^{14}$ cm^{-3}, but low end of range difficult to achieve
Range of dopants available	A wide range, most groups II, IV and VI elements can be used. Limited only by solubility in liquid metal and should have a low vapor pressure over the solution at the growth temperature	A very wide range of II, IV and VI elements. Limited only by need for solubility of element or a decomposable compound	A rather narrow choice, Si, Sn (n-type) Be, Mn (Mg) p-type
Control of dopant incorporation	Range available $10^{14}–10^{19}$ cm^{-3}, but only flat profiles can be produced, i.e., sudden changes in doping level cannot be obtained	Range available $5 \cdot 10^{14}–10^{19}$ cm^{-3} and reasonably sharp changes in dopant concentration (spatial resolution $\approx$30 nm)	Range available $5 \cdot 10^{16} – 10^{19}$ cm^{-3}. With sharp changes in dopant concentration (spatial resolution $\approx$5 nm)
Process amenable to automation	Probably not, too operator dependent	To some extent, but some process steps would still be operator dependent	The whole process can be automated.

4.1.4   Ion Implantation

The Accelerator

Ion implantation is a doping technique with which ions are shot into a substrate (e.g., a silicon wafer) using an accelerator. The basic principle is presented in Fig. 4.16.

The desired ion species is let in as a gaseous compound through a needle valve (alternatively, solid state sputtering sources are used). The compound is dissociated and ionized with an electron beam. The arising ions (including the unwanted ones) are pulled out of the source area and brought to an energy of 30 keV by a first, preliminary acceleration (all figures are typical values). Then, the ions pass through a magnetic field filter which is adjusted in such a way that only the desired ion type can run into the acceleration tube. The magnetic field filter is based on the fact that the Lorentz force for a moving charged particle compels a circular path. The radius of the circle depends on the magnetic field B, the velocity v, and the mass m of the particle. More exactly, it depends on m/e (e is the elementary charge). For the desired ion species, i.e., for a given m/e, the magnetic field is adjusted in such a way that the circular path of these particles terminates exactly at the end of the accelerator tube. There, the ions acquire a total energy of 360 keV. This energy can be doubled or multiplied by the use of double or multiple charged ions. However, the ion yield, i.e., the available ion current, is exponentially reduced with the state of charge (ionization state).

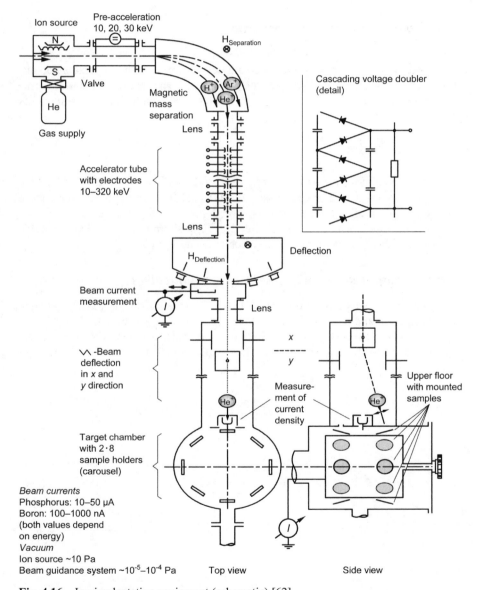

Fig. 4.16   Ion implantation equipment (schematic) [62]

The beam can be positioned by a combination of an aperture with a Faraday cage. With two capacitor disks each, the beam can be scanned upwards and downwards or left and right. In order to avoid Lissajou figures, horizontally and vertically incommensurable scanning frequencies are chosen. The beam current is measured by an ammeter, which is connected to the substrate holder isolated against ground. The substrate holder is designed as a carousel, for example, in order to be able to implant several samples without an intermediate ventilation.

Calculation of the Implantation Time

Typical required dose values N_I lie between 10^{12} and 10^{16} cm^2. The necessary implantation time depends on the available beam current, I, on the irradiated substrate surface, A, and on the charge state of the ions. The ions summed in the irradiation time t represent a charge $Q = q\,N_I\,A$, whose relation to the ion beam results in the irradiation time:

$$t = \frac{Q}{I} = \frac{q\,N_I\,A}{I}\,. \qquad (4.1)$$

As an example, an ion current of 1 µA, a substrate surface of 100 cm^2, and a dose of 10^{13} cm^{-2} deliver an irradiation time of 160 s. This result applies to the charge state 1 (simple ionization) of the ions. For double, triple, etc. charged ions, the implantation time must be doubled (tripled, etc.). It should be noted that for multiple charged ions an equivalent multiple current is measured.

Lattice Incorporation, Radiation Damage, Annealing

The penetrating ions pass through the lattice depending on the ion mass/lattice atom mass ratio and the momentary velocity in a zigzag path (Fig. 4.17)

At the jags of the path, the ions impinge on host lattice atoms, which leave their places and go to the interstitial site. Thus, lattice defects of different nature develop.

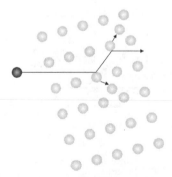

Fig. 4.17   Path of an implanted ion in a lattice

The simplest defect—the Frenkel defect—is produced by the displacement of a lattice atom into an interstitial site. Thus, a vacancy and an interstitial atom develop. Vacancies can possess different charge states (e.g., neutral, positive, negative, double negative). Furthermore, they can form aggregates with foreign atoms and influence their diffusion. Double vacancies can be formed if an impinging ion dislodges two nearest neighbor lattice atoms. Moreover, they can be formed by two single vacancies. Double vacancies are stable up to approximately 500 K.

Dislocations can develop by the association of single defects, or they grow from unannealed radiation damage into undamaged area during annealing. Dislocation lines anneal only at high temperatures (≥1000 °C), and very often they do not anneal at all in implanted layers.

Further defects can be formed by the accumulation of vacancies and interstitial atoms as well as by the association of foreign atoms with vacancies or interstitial atoms.

If many lattice atoms are displaced by an impact ion in a considerably small volume, a locally amorphized area develops. Often, this area is known as cluster, and its exact structure is unknown. In ion implantation, one always expects this case because of the high mass and energy of the impact particles. Accordingly, the possible processes during implantation and subsequent temperature annealing are complex and hardly accessible to theoretical description.

The implanted ion comes to rest usually on an interstitial site after numerous impacts. Thus, contrary to intention, it cannot work as a dopant (Recall: the 15th electron of phosphorus can only be detached with small energy expenditure by embedding the atom into the lattice structure and emerge as a free electron in the crystal).

Annealing is applied for both the annealing of lattice defects and for the relocation of the doping atoms from interstitial sites into lattice sites, i.e., a thermal treatment of the implanted samples at approximately 900–1100 °C in a suitable atmosphere like nitrogen or hydrogen gas. The ratio of electrically active ions sitting on lattice sites to the total number of implanted ions is called activation. Usually, the activation rises monotonously with temperature. In the case of phosphorus, for example, almost complete activation is achieved at approximately 700 °C. However, there are exceptions like in the case of boron where an intermediate minimum can occur (Fig. 4.18). The intermediate minimum of the boron implantation is caused by the behavior of the interstitial silicon atoms, which are produced by the nuclear impacts during the implantation. At approximately 500 °C, they try to return to lattice sites thereby pushing the already existing boron atoms in lattice sites back to interstitial sites.

Implantation Profile

To a first approximation, the distribution of the implants is described by a Gaussian curve:

$$N(x) = N_{max}\, e^{-\left(\frac{x-R_p}{\Delta R_p \sqrt{2}}\right)^2}.$$

(4.2)

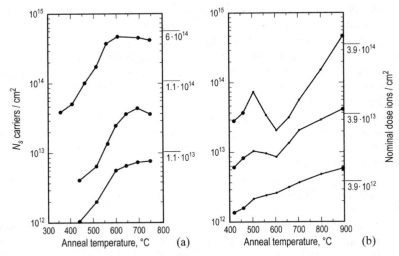

Fig. 4.18 (a) Electrically active phosphorus atoms in Si. Dose values are $1.1 \cdot 10^{13}$, $1.1 \cdot 10^{14}$, and $6 \cdot 10^{14}$ cm^{-2}, implantation energies 20 and 40 keV, (b) electrically active phosphorus atoms in Si. Dose values are $1 \cdot 10^{13}$, $1 \cdot 10^{14}$, and $1 \cdot 10^{15}$ cm^{-2}, implantation energies 20, 30, and 50 keV. According to [63], slightly modified

R_p is the position of the center of the distribution (for the Gaussian curve, this is identical to the maximum position), ΔR_p the full width at half maximum of the distribution (Fig. 4.19).

$N(x)$ has the of dimension of cm^{-3}. The concentration, N_{max}, in the maximum of the distribution is calculated from the implanted doses N_I, $[N_I] = $ cm^{-2}:

$$\int_0^\infty N(x)\, \mathrm{d}x = N_I \tag{4.3}$$

or

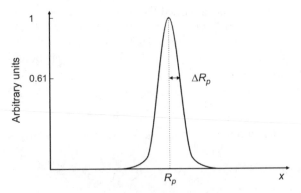

Fig. 4.19 Simplified distribution of the implants in the substrate

$$N_{\max} \cong 0.4 N_I / \Delta R_p .\tag{4.4}$$

As a prerequisite for the validity of Eq. 4.2, the implantation profile must not diffuse by annealing. A second prerequisite is the prevention of an implantation into a low indexed crystal orientation. Thus, the crystal must act amorphous for the ion beam. If one shoots toward a low indexed crystal orientation, then the ion beam runs without resistance through the lattice channels and reaches substantially large depths *(channeling)*, Fig. 4.20.

Channeling is avoided by tilting the crystal, which is aligned in (100) direction against the ion beam, usually by $7°$ as depicted in Fig. 4.21.

The Gaussian function used in Eq. 4.2 is only an approximation of the distribution predicted by theory. This "exact" distribution cannot be expressed by an analytic function. Rather, the ion profile can be subjected to the so-called "moment development". This procedure can be roughly compared to a series expansion of a function for Fourier coefficients or a multipole development: the ion profile $N(x)$ is multiplied by 1, x, x^2, etc. and integrated over the entire semiconductor depth x. The function $f(x)$ can be reconstructed from the developing numerical values (the moments). These numerical values arise from transport-theoretical considerations.

At least the first four moments can be illustrated: $\int N(x)\,dx$ describes the zero moment identical to the implanted dose, $\int x\,N(x)\,dx$ describes the first (static) moment of the center of the distribution, $\int x^2\,N(x)\,dx$ delivers its straggling and $\int x^3\,N(x)\,dx$ its skewness (asymmetry). Just from these descriptions alone, the Gaussian function is obviously attractive for describing the profiles: from the above integrations, this function delivers R_p as a center, ΔR_p as full width at half maximum, and zero as skewness.

In all technically important cases, it is sufficient to describe skewed profiles by joining two Gaussian functions with the half widths σ_1 on the right and σ_2 on the left of the maximum. Let R_m denote the position of the maximum; R_p is only equal

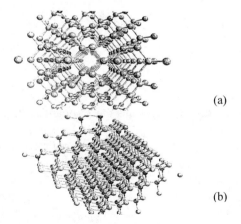

(a)

(b)

Fig. 4.20   (a) Si crystal in (110) direction and (b) misaligned by $7°$

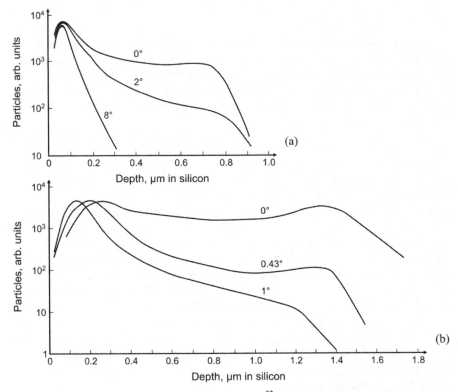

Fig. 4.21 Dependency of the depth distribution of ^{32}P after 40 keV (a) and 100 keV (b) implantation. The angles indicate the deviations from the (110) direction [64].

to R_m when $\sigma_1 = \sigma_2$. Thus,

$$N(x) = \frac{2}{(\sigma_1 + \sigma_2)\sqrt{2\pi}}\ e^{\frac{(x-R_m)^2}{2\sigma_1^2}} \qquad \text{for } x \geq R_m, \text{ and}$$

$$\hspace{6cm} (4.5)$$

$$N(x) = \frac{2}{(\sigma_1 + \sigma_2)\sqrt{2\pi}}\ e^{\frac{(x-R_m)^2}{2\sigma_2^2}} \qquad \text{for } x \leq R_m.$$

In reference tables, however, the energy R_p, ΔR_p, and the standardized third moment $CM_{3p}/(\Delta R_p)^3$ are listed (Table 4.2).

The way of getting back from this data to σ_1, σ_2, and R_m is complicated. The equation

$$\frac{CM_{3p}}{\sigma_p^3} = 2\frac{\Delta}{\sigma_p}\left[0.8 - 0.256\left(\frac{\Delta}{\sigma_p}\right)^2\right] \qquad (4.6)$$

with known $CM_{3p}/(\sigma_p)^3$ and σ_p is solved for Δ/σ_p and subsequently $\Delta = (\sigma_1 -$

$\sigma_2)/2$ is obtained. With this background, the second equation

$$1 = \left(\frac{\sigma_m}{\sigma_p}\right)^2 + 0.44\left(\frac{\Delta}{\sigma_p}\right)^2 \qquad (4.7)$$

is solved for $\sigma_m = (\sigma_1 + \sigma_2)/2$, and thus σ_1 and σ_2 are given. Finally, R_m is given by:

$$R_m = R_p - 0.8(\sigma_2 - \sigma_1). \qquad (4.8)$$

If the third moment is negative, σ_1 and σ_2 must interchange their roles.

4.1.5   Formation of Silicon Oxide

Thin oxide layers are contained in almost all electronic devices. They appear as gate oxide in MOS transistors or MIS solar cells, field oxide for isolation purposes, anti-reflection layers in solar cells, or as passivation layers for long-term protection.

We begin with thermal oxidation. A Si wafer is cleaned so that any organic or heavy metal impurity on the surface is removed and the natural oxide is dissolved. Then, the wafer is inserted in a quartz tube heated to a temperature of about 1100 °C. A flow of an oxidizing gas, either pure oxygen *(dry oxidation)* or nitrogen driven through water *(wet oxidation)*, is maintained. When oxygen penetrates into the substrate, the Si surface reacts with the oxygen and forms silicon dioxide. While this procedure sounds simple, it requires highest cleanliness, which is the critical step for the MOS production. Figures 4.22 and 4.23 show graphs of the oxide thickness *vs.* oxidation time for several temperatures [66].

Numerous scientific and technical investigations have focused on the properties of silicon dioxide. Fields of interest are the growth laws, deep levels, capture of charge carriers from the silicon, segregation and rearrangement of the dopant in the neighboring silicon, masking properties against diffusion and ion implantation, etc. Their in-depth discussion is beyond the scope of this book.

Technical alternatives to thermal oxidation are CVD and PECVD of oxides. These are treated in the section on CVD. There are some technical CVD versions such as TEOS deposition shown in Fig. 4.24.

A feed gas (usually nitrogen) is driven through a container filled with tetra-ethylorthosilicate (TEOS). TEOS is a liquid at room temperature. Its chemical structure is presented in Fig. 4.25.

The enriched nitrogen flows to the wafers where SiO_2 is deposited on their surfaces. The deposition is maintained at a temperature of about 650 to 850 °C by means of an external induction coil. It should be noted that the silicon in the TEOS is already oxidized, in contrast to the silane process.

In this state, TEOS finds only limited application because the deposition temperature (>650 °C) prevents its use after metallization. In order to obtain lower deposition temperatures, the application of a more aggressive oxidant, i.e., ozone, is required. After adding some few molar per cent of ozone, the optimum deposition temperature is reduced to 400 °C.

Table 4.2   Ranges, standard deviation, and third moment (beside other parameters) for boron implantation with energies of 10 keV to 1 MeV [65]

LSS range statistics for boron in silicon.
Substrate parameters Si: $Z = 14$, $M = 28.090$, $N = 0.4994 \cdot 10^{23}$, $\rho/r = 0.3190 \cdot 10^{2}$,
$\quad\quad \varepsilon/e = 0.1130$, CNSE $= 0.3242 \cdot 10^{2}$, $\mu = 2.554$, $\gamma = 0.8089$, SNO $= 0.9211 \cdot 10^{2}$
Ion B: $Z = 5$, $M = 11.000$
Northcliffe constant $= 0.292 \cdot 10^{4}$

Energy, keV	Projected range, µm	Projected standard deviation, µm	Third moment ratio estimate	Lateral standard deviation, µm
10	0.0333	0.0171	-0.031	0.0236
20	0.0662	0.0283	-0.309	0.0409
30	0.0987	0.0371	-0.483	0.0555
40	0.1302	0.0443	-0.617	0.0682
50	0.1608	0.0504	-0.727	0.0793
60	0.1903	0.0556	-0.821	0.0891
70	0.2188	0.0601	-0.904	0.0980
80	0.2465	0.0641	-0.978	0.1061
90	0.2733	0.0677	-1.046	0.1135
100	0.2994	0.0710	-1.108	0.1203
110	0.3248	0.0739	-1.166	0.1266
120	0.3496	0.0766	-1.220	0.1325
130	0.3737	0.0790	-1.271	0.1380
140	0.3974	0.0813	-1.319	0.1431
150	0.4205	0.0834	-1.364	0.1480
160	0.4432	0.0854	-1.408	0.1525
170	0.4654	0.0872	-1.449	0.1569
180	0.4872	0.0890	-1.489	0.1610
190	0.5086	0.0906	-1.527	0.1649
200	0.5297	0.0921	-1.564	0.1687
220	0.5708	0.0950	-1.634	0.1757
240	0.6108	0.0975	-1.699	0.1821
260	0.6496	0.0999	-1.761	0.1880
280	0.6875	0.1020	-1.820	0.1936
300	0.7245	0.1040	-1.876	0.1988
320	0.7607	0.1059	-1.930	0.2036
340	0.7962	0.1076	-1.981	0.2082
360	0.8309	0.1092	-2.030	0.2125
380	0.8651	0.1107	-2.078	0.2166
400	0.8987	0.1121	-2.125	0.2205
420	0.9317	0.1134	-2.170	0.2242
440	0.9642	0.1147	-2.214	0.2277
460	0.9963	0.1159	-2.257	0.2311
480	1.0280	0.1171	-2.298	0.2344
500	1.0592	0.1182	-2.339	0.2375
550	1.1356	0.1207	-2.435	0.2448
600	1.2100	0.1230	-2.526	0.2515
650	1.2826	0.1252	-2.614	0.2576
700	1.3537	0.1271	-2.697	0.2633
750	1.4233	0.1289	-2.778	0.2687
800	1.4917	0.1306	-2.856	0.2737
850	1.5591	0.1322	-2.933	0.2784
900	1.6254	0.1337	-3.006	0.2829
950	1.6909	0.1351	-3.079	0.2871
1000	1.7556	0.1364	-3.149	0.2912

Table 4.2 (cont'd)   Ranges, standard deviation, and third moment (beside other parameters) for boron implantation with energies of 10 keV to 1 MeV [65]

Energy, keV	Range, µm	Standard deviation, µm	Nuclear energy loss, keV / µm	Electronic energy loss, keV / µm
10	0.0623	0.0141	96.86	102.2
20	0.1100	0.0221	75.89	144.0
30	0.1536	0.0276	63.09	175.8
40	0.1940	0.0316	54.43	202.3
50	0.2317	0.0347	48.12	225.4
60	0.2673	0.0371	43.28	246.1
70	0.3010	0.0392	39.44	264.9
80	0.3331	0.0409	36.30	282.2
90	0.3638	0.0424	33.68	298.4
100	0.3934	0.0437	31.45	313.4
110	0.4218	0.0449	29.52	327.7
120	0.4494	0.0459	27.85	341.1
130	0.4761	0.0468	26.37	353.9
140	0.5020	0.0476	25.05	366.0
150	0.5272	0.0484	23.88	377.6
160	0.5518	0.0491	22.81	388.7
170	0.5759	0.0497	21.85	399.4
180	0.5993	0.0503	20.97	409.7
190	0.6223	0.0509	20.17	419.5
200	0.6448	0.0514	19.43	429.1
220	0.6886	0.0523	18.11	447.1
240	0.7309	0.0532	16.97	464.0
260	0.7718	0.0539	15.98	479.9
280	0.8116	0.0546	15.10	494.9
300	0.8503	0.0552	14.32	509.1
320	0.8880	0.0558	13.62	522.6
340	0.9249	0.0564	12.99	535.4
360	0.9610	0.0569	12.42	547.5
380	0.9964	0.0573	11.90	559.1
400	1.0311	0.0578	11.42	570.2
420	1.0651	0.0582	10.98	580.7
440	1.0987	0.0586	10.58	590.9
460	1.1317	0.0589	10.20	600.6
480	1.1642	0.0593	9.854	609.9
500	1.1962	0.0596	9.530	618.8
550	1.2745	0.0604	8.809	639.6
600	1.3506	0.0611	8.192	658.6
650	1.4246	0.0618	7.659	675.8
700	1.4969	0.0624	7.193	691.7
750	1.5678	0.0629	6.781	706.2
800	1.6373	0.0634	6.415	719.5
850	1.7056	0.0639	6.088	731.8
900	1.7728	0.0644	5.792	743.2
950	1.8391	0.0648	5.525	753.6
1000	1.9046	0.0653	5.282	763.3

The anodic oxidation [67] is shown in Fig. 4.26. The wafer is immersed into a 0.04 M solution of KNO_3 in ethylene glycol with a small addition of water. After mounting it to a holder with a vacuum, it is positively charged, while a platinum disk acts as a backplate electrode.

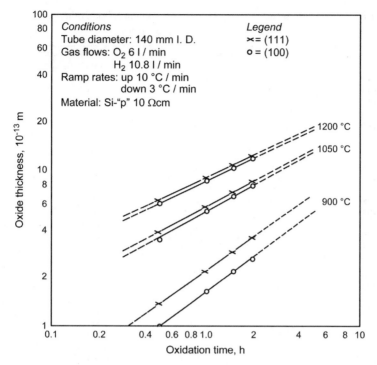

Fig. 4.22   Oxide thickness *vs.* oxidation time (wet oxidation)

The current causes a reaction on the surface of the silicon:

$$Si + 2H_2O + 4h^+ \rightarrow SiO_2 + 4H^+ \tag{4.9}$$

$$CH_2OH - CH_2OH \rightarrow 2HCHO + 2H^+ + 2e^- \tag{4.10}$$

$$H_2O \rightarrow \tfrac{1}{2}O_2 + 2H^+ + 2e^- \tag{4.11}$$

Technically, anodic oxidation is of little importance. The quality of the oxide is too low, and the process is time-consuming or not compatible with other applications.

Rarely, sol gels are used to manufacture oxide layers. During this process, a suspension of oxide particles in an organic solvent is distributed over the wafer. A centrifuge facilitates homogeneous distribution of the liquid on the wafer. Then, the wafer is baked and the solvent evaporates. The required temperatures range from 500 to 800 °C. The oxide can have the quality of a gate oxide, but no large-scale application has been reported so far.

A further manufacturing process is SOI (silicon-on-oxide), i.e., the implantation of oxygen up to the stoichiometric dose and its reaction with silicon to form SiO_2 (see Sect. 4.1.4 on ion implantation). SOI has gained in importance because of better radiation resistance and heat distribution compared to conventional technologies.

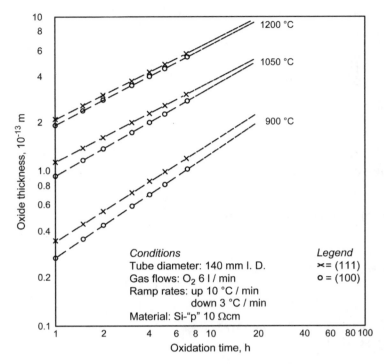

Fig. 4.23 Oxide thickness *vs.* oxidation time (dry oxidation)

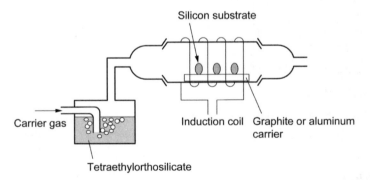

Fig. 4.24 TEOS process

4.2 Characterization of Nanolayers

4.2.1 Thickness, Surface Roughness

In the following chapter, some methods to determine nanometer-thin layers are presented. They are demonstrated with the help of measurement setups.

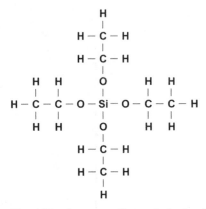

Fig. 4.25 Structure of tetraethylorthosilicate (TEOS)

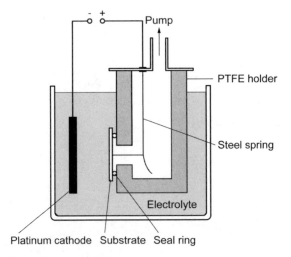

Platinum cathode Substrate Seal ring

Fig. 4.26 Assembly of the anodic oxidation of silicon (schematic)

Admittance Bridges

A metal oxide semiconductor (MOS) diode is produced by the oxidation of a Si wafer with, for instance, an oxide of 10 nm thickness and metal evaporation on the oxide (this structure being the middle part of the so-called MOS transistor). From the specifications of the evaporation (or the lithography), area A of the metal point is known. The measurement of the oxide thickness is performed with the help of an admittance (or capacitance) bridge (Fig. 4.27). Additionally, the substrate is operated in accumulation by applying a bias voltage (attraction of majority carriers to the silicon boundary surface). The modulating frequency of the bridge should be as low as possible in order to suppress the resistance of the silicon bulk. A good choice is 200 Hz. A detailed description is given in the section on the determination of the doping level and profile (Sect. 4.2.4.)

The measured capacitance is caused by the oxide alone. Therefore, it is called C_{ox}. Now the equation for the parallel-plate capacitor can be applied to give the oxide thickness:

$$d_{ox} = \varepsilon_{ox} \frac{A}{C_{ox}}$$
(4.12)

where A is the area of the metal point, and $\varepsilon_{ox} = 3.4 \cdot 10^{-13}$ F / cm.

Michelson Interferometer Measurements

It is presupposed that the film can be etched locally in such a way that (i) a sharp edge develops, and (ii) the exact thickness of the film to be measured is removed at the etched position. This case is frequently implemented, for instance, if SiO_2 is etched by HF with high selectivity against Si (i.e., the etching stops when the Si is reached). Another case is a thin layer of lightly-doped silicon on a carrier of heavily doped silicon. The surface, which has a step, is made reflective by metal evaporation and inserted into a Michelson interferometer as shown in Fig. 4.28.

Two parallel sets of bright and dark interference fringes appear to the viewer. They are shifted against each other by ΔN. The height of the step and thus the thickness of the film d is given by:

$$d = \Delta N \frac{\lambda}{2} .$$
(4.13)

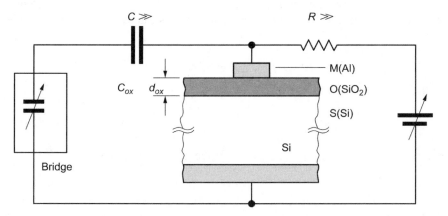

Fig. 4.27 Measurement of the oxide layer with an admittance bridge

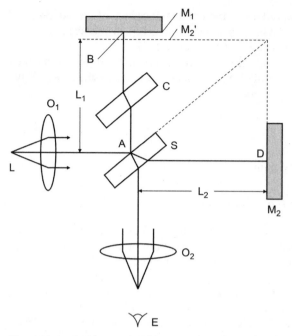

Fig. 4.28 Michelson interferometer [68]. L is the light source, M_1 and M_2 are mirrors, C is a disk for the compensation of the double way A–D–A by the beam splitter S. M_2' is the virtual image of M_2. O_1 and O_2 are the objective lenses.

Tolanski Method

This method is referred to as a special case of a multiple-beam interference. The method can be explained in two steps:

(i) *Multiple beam interference* (Fig. 4.29). After passing through a glass plate, each beam is partially reflected *(R)* and partially transmitted *(T)* because of a thin silver film. According to Airy's formula, the intensities can be described by:

$$I_T = \frac{I_{max}}{1 + F\ \sin^2(k\ \delta/2)} \tag{4.14}$$

with the phase difference

$$k\ \delta = \frac{2\pi}{\lambda}\ 2n_1\ t\ \cos\theta = 2\pi N\ , \tag{4.15}$$

the finesse (i.e., the separation of adjacent fringes per width of half max)

$$F = \frac{4R}{(1-R)^2}\ , \tag{4.16}$$

and

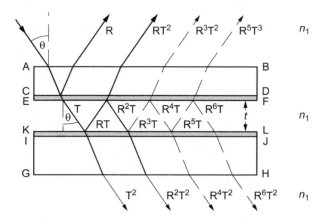

Fig. 4.29   Multiple-beam interference (modified from [68] and [69])

$$I_{max} = \frac{T^2}{(T + A)^2} \, .$$ (4.17)

n_1: the refractive index of the medium between the mirror glasses.
A: absorption of the mirrors DF and LJ
T: transmission of the mirror
R: reflection of the mirror.

With a sufficient reflection, R ($\cong 95\,\%$), transmitted light becomes narrow bright fringes on a dark background if irradiation is done with monochromatic light. In reflection, narrow dark fringes on a bright background are seen. There, the prerequisite of constructive interference is fulfilled:

$$N = 1, 2, 3...$$ (4.18)

(ii) *Fizeau strips.* The sample preparation resembles that of the above-described Michelson experiment. The sample to be measured is prepared in such a way that a part of the film is etched or the substrate is already partly exposed during deposition (Fig. 4.30a). Now it used instead of the lower glass plate of Fig. 4.29, and the upper glass plate is tilted against the sample at a small angle.

When illuminating with a monochromatic (point) source, an interference pattern is obtained for interferences of the same thickness. More precisely, two interference fringes shifted against each other develop again. Their misalignment, ΔN (which can also be a fractional number), yields the film thickness:

$$d = \Delta N \frac{\lambda}{2} \, .$$ (4.19)

If the interference patterns of the etched side lie exactly between the strips of the unetched side, $\Delta N = \frac{1}{2}$ and the film thickness $d = \lambda / 4$. A result is depicted in Fig. 4.31.

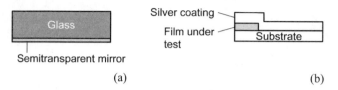

Fig. 4.30   Sample preparation for the Tolanski method: (a) Fizeau plate, (b) sample

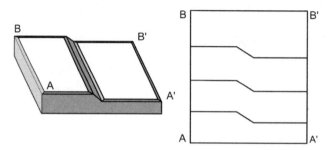

Fig. 4.31   Tolanski measurement (schematic)

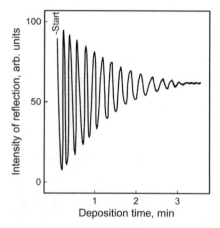

Fig. 4.32   Interference patterns of the emitted infrared radiation by 0.55 μm polycrystalline silicon deposition on Si_3N_4 [70, slightly changed]. Reactant: SiH_2Cl_2–H_2. Deposition temperature: 1120 °C. Detector: PbS—2.0 μm filter. Evidence: infrared radiation

In Situ Method

An example of the measurement of the layer thickness during the deposition of polycrystalline silicon on silicon nitride is given in Fig. 4.32.

By flowing $SiCl_4/H_2$ or SiH_2Cl_2/H_2 gas over a Si_3N_4 substrate, polycrystalline silicon is deposited on the substrate by the dissociation of the silane. Due to the high temperature of 1120 °C, the sample emits infrared light, and selected frequencies can be observed (here 2.0 µm). Because of multiple interference in the film, an interference pattern develops. With otherwise constant parameters (wavelength, observation direction), the interference changes continuously between constructive and destructive interference with progressive growth. The same interference patterns are obtained if an external monochromatic source is used.

Frequency Shift of a Quartz Crystal Oscillator

During the deposition of metallic films, a quartz resonator is placed in the proximity of the material that is to be deposited. This quartz reduces its resonance frequency with the thickness of the film deposited on it. The frequency change is converted into units of the layer thickness.

This procedure is not very exact. Errors are caused by the different distances of resonator/deposition source and substrate/deposition source. Moreover, the unknown initial treatment of the quartz leads to an irreproducible behavior. A separate correction factor must therefore be determined for each metal.

Ellipsometry

As a rule, the thickness of either one or two transparent films on an opaque substrate is measured with the ellipsometer. An elliptically polarized monochromatic beam impinges on the surface of the film (Fig. 4.33).

The incident wave is characterized by the amplitudes of the parallel (see below) and the perpendicular (see below) components and the phase difference Δ between these two components. After reflection at both surfaces and the summation of all beams, the relation between the amplitudes and the phase difference is modified. This modification depends on

- the refractive indices n_2, n_3 of film and substrate,
- the absorption coefficients k_2, k_3 of film and substrate,
- the incidence angle φ_1, and
- the film thickness d.

On condition that the optical constants n_3 and k_3 are known and that the film is transparent, i.e., $k_2 = 0$, the refractive index and the thickness of the film are determined from a quite simple experiment (Fig. 4.34).

Monochromatic light is first linearly polarized by a polarizer, and then elliptically polarized by a retarder (quarter wave retarder). After reflection at the film, the light passes through a further polarizer for analysis and is subsequently observed with a telescope.

For didactic purposes, let us exchange the sequence of the compensator and the film. Additionally, let us define a reference plane, for instance, the plane of incidence, E. It contains the incident and outcoming beams and the perpendicular. The terms *parallel* and *perpendicular* refer to this plane, furthermore to the numerical angles of the quarter wave retarder, the polarizer, and the analyzer. The polarizer (in an arbitrary position) produces a linearly polarized oscillation, which may be

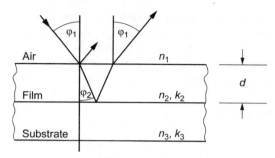

Fig. 4.33   Geometric and material definitions in the ellipsometer experiment

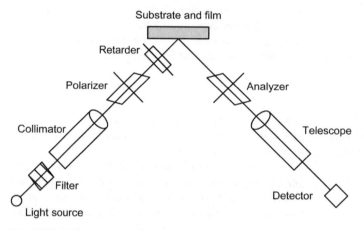

Fig. 4.34   Ellipsometer [71]

split up into a parallel and a perpendicular fraction. The two fractions have a phase difference $\Delta = 0$. After reflection the new phase difference Δ of the two fractions transforms a linear oscillation into an elliptical oscillation of the form:

$$\frac{x^2}{A^2} + \frac{y^2}{B^2} - \frac{2x\,y\,\cos\Delta}{A\,B} = \sin^2\Delta \,.$$ (4.20)

A and B are the amplitudes of the two oscillation directions in the previous coordinate system (Fig. 4.35). It should be noted that the phase angle Δ between the components of the ellipse applies to the selected coordinate system only. By rotating the coordinate system, the phase changes. In particular, $\Delta = \pi/2$ holds if the large semiaxis and the new x' axis coincide. Such a coordinate transformation takes place if the elliptically polarized beam is exposed to a quarter wave retarder. Its two polarization directions are put in the directions of the ellipse axis so that the fast beam corresponds to the lagging component of the ellipse (angle of rotation ϑ of the quarter wave retarder from the x-direction). Since the two beams

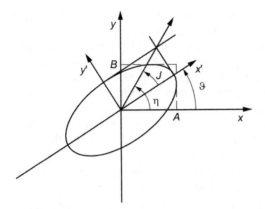

Fig. 4.35 Oscillation ellipse after reflection

catch up with each other again, a linearly polarized beam develops, which can be brought to extinction with an analyzer. The direction of the linearly polarized beam to the reference plane is η. η (rather: $\eta + \pi/2$) is measured with the analyzer, ϑ is known as the direction of rotation of the quarter wave retarder. The criterion of the correct determination of η and ϑ is the extinction of the beam behind the analyzer. Thus, the axes ratio of the ellipse is given as

$$\tan J = \tan(\eta - \vartheta). \tag{4.21}$$

The knowledge of this angle enables us to determine the ratio of the oscillation components $B/A = \tan \psi$ in the reference plane system:

$$\cos 2\psi = \cos 2J \cos 2\vartheta. \tag{4.22}$$

For the phase difference Δ in the reference plane system, we find

$$\tan \Delta = \frac{\tan 2J}{\tan 2\delta}. \tag{4.23}$$

What has been achieved so far? Originally, a linearly polarized beam with the components A_0 (parallel to the reference or incidence plane) and B_0 (perpendicular to it) is produced in the polarizer. This beam has the phase difference $\Delta_0 = 0$ (otherwise it would not be linearly polarized) and the amplitude ratio $\psi_0 = B_0/A_0$, which is known from the rotation of the polarizer against the reference plane by the angle ψ_0. A particularly simple solution is obtained in the case of a 45° rotation; $A_0 = B_0$, or $\tan \psi_0 = 1$.

This method enables us to determine how the amplitude ratio and the phase difference have changed due to reflection. If a theoretical statement can be made on the ψ-Δ change resulting from the reflection at a thin film of a thickness d and refractive index n_2 [i.e., $\tan \psi / \tan \psi_0 = f(d, n_2)$ and $(\Delta - \Delta_0 = f(d, n_2)$)], then the determination of the layer thickness and refractive index should be possible by forming the inverse functions $d(\psi, \Delta)$ and $n_2(\psi, \Delta)$.

The theory of Fresnel and Neumann offers a solution. Its knowledge, however, is not required for the operation of the ellipsometer, as will be shown below.

For historical reasons, the procedure is done somewhat differently without changing the basic idea. First, the compensator is usually set behind the reflecting surface. Therefore, the sequence is polarizer–sample–compensator–analyzer. Second, one does not rotate the compensator but the polarizer (for the compensator, a direction of ±45° of the fast axis is maintained against the reference plane). Thus, by varying the amplitude ratio B_0/A_0, the ellipse created after reflection is rotated. If the position of the compensator is adjusted correctly, the elliptical oscillation is brought back into a linear form. If the angles ψ and Δ are determined in this way, then these two values can be brought into a set of curves where Δ is presented as a function of ψ. The parameters are film thickness (in units of wavelength or as phase difference $2\pi n_2\, d/\lambda$) along a Δ-ψ curve and the refractive index which has a fixed value for each Δ-ψ curve. An example of such a curve set is found Fig. 4.36.

These curves are calculated according to the Fresnel Neumann theory. For each wavelength, angle of inclination of the two polarization levers against the sample and path difference of the compensator, a new record of curves must be calculated. In our graph, many specializations have taken place. For instance, the quarter wave retarders can be replaced by a compensator of any path difference. The analysis must then be corrected accordingly. Moreover, we have not treated the other pairs of solutions for the compensator and analyzer angle. Furthermore, an accurate discussion about the determination of the direction of rotation of the ellipse must be done. This is connected with the question about the sign of the directions of rotation during the measurement and the trigonometric functions. At last, we neglected the question of how the absorption influences the measurement.

It should be noted that the ellipsometer can easily be automated by a microprocessor controller. Of course, a technique is required that enables us to deter-

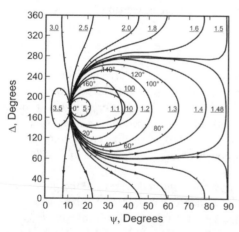

Fig. 4.36 Δ-ψ curve set. Parameters are phase angle $2\pi n_2\, d/\lambda$ (along a curve) and n_2 (from curve to curve); normal angle of the ellipsometer levers $\varphi_1 = 70°$, $n_3 = 4.05$ (silicon), extinction coefficient of the silicon $k_3 = 0.028$, $\lambda = 546.1$ nm [71]

mine the two angles of rotation of analyzer and polarizer simultaneously. Such devices are commercially available.

A completely different approach consists of measuring the ellipse by rotating the analyzer photometrically (with the multiplier). Such devices have been constructed as well.

In the above discussion, a homogeneous layer is assumed whose refractive index is constant with depth. Two values Δ and ψ are measured, which deliver the thickness and the refractive index. However, our assumption can be invalid if, for instance, one or more films are deposited on the first one, or if the conditions of the depositions are changed so that a refractive index profile is produced. Therefore, an improved model and a sophisticated instrumentation is required. Instead of using a single ellipsometer wavelength, the whole available spectrum can be used. This method is known as spectroscopic ellipsometry, which requires new approaches in data evaluation. On the basis of well known technological data such as film thickness, film material, and profiles, a model of the layer structure can be set up and the spectral Δ-ψ curves can be determined. The model is subsequently improved by fitting the resulting Δ-ψ curves to the measured curves. The knowledge on some layers is so good that the derived profile even delivers physical models of the film. As an example, the ratio of amorphous to microcrystalline fractions can be determined during growth of amorphous films. The surface roughness of the substrate can be determined, and the transition layers between substrate and film or between the films can be resolved to within Ångströms.

In spectroscopic ellipsometry, the components of the dielectric function ε rather than Δ and ψ are plotted. The transformation of Δ and ψ to ε is given by

$$\varepsilon_1 = \sin^2 \varphi_1 \left(1 + \frac{\tan^2 \varphi_1 (\cos^2 2\psi - \sin^2 2\psi \sin^2 \Delta)}{(1 + \sin 2\psi \cos \Delta)^2} \right) \text{ and} \tag{4.24a}$$

$$\varepsilon_2 = \frac{\sin^2 \varphi_1 \tan^2 \varphi_1 \sin 4\psi \sin \Delta}{(1 + \sin 2\psi \cos \Delta)^2}. \tag{4.24b}$$

An example of the dielectric function (ε_2) depending on the crystalline state of Si is shown in Fig. 4.37. An evaluation of the dielectric function and the transformation to a layer model is depicted in Fig. 4.38.

Profilometer

A thin film, whose thickness is to be measured, is grown on a substrate. Therefore, the sample is partly covered with wax or photoresist so that a sharp edge between the covered and untreated surface is given. The untreated surface is etched in acid, which removes the film but does not attack the substrate (selective etching). Consequently, the edge deepens in the sample. Now, the cover (wax) is removed. The edge's depth is measured with a needle, which is moved across the edge and which is sensitive to changes in the surface. (Fig. 4.39).

The vertical position of the needle is checked with a piezoconverter (with step heights larger than 2 nm) or an inductive converter (with step heights larger than

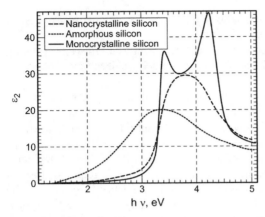

Fig. 4.37   Dielectric functions for amorphous, polycrystalline, and monocrystalline Si [72]

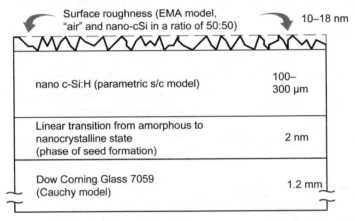

Fig. 4.38   Layer sequence derived from spectroscopic ellipsometry [72]

some micrometers). These machines are called Talysurf or Talystep. An example of a measurement is shown in Fig. 4.40. When small edge differences are to be measured, the major difficulty in the handling of the device is the leveling of the unetched surface serving as measuring reference. If this plane and the needle's path do not correspond, the noise during the high measurement amplification will prevent a reasonable determination of the height. In the meantime, however, self-adjusting versions are available.

Scanning Tunneling Microscopy (STM), Atomic Force Microscopy (AFM)

- STM. The surface of a conducting material is covered with an electron cloud, whose density reduces with increasing distance from the surface.

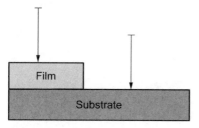

Fig. 4.39   Thickness measurement with a needle

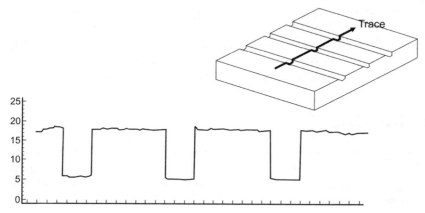

Fig. 4.40   Measuring an edge with an inductive needle [73]. Quartz deposition on glass substrate. Test ridges are produced by removing the mask. Vertical magnification 1,000,000 fold, one small division represents 2 nm. Horizontal magnification 200fold, one small division represents 0.025 mm. Thickness of the deposited layer (mean value) approximately 26 nm (25.9 mm on the diagram)

If a metal tip is within a close distance to this surface, a current flows between the tip and the surface. The current flow begins from a distance of about 1 nm, and it decreases by a factor of 10 for every reduction of 0.1 nm in the distance. This phenomenon can be used for an x-y presentation of the roughness. When moving the tip laterally, a constant current is maintained by following the distance of the tip. The necessary adjustment is a measure of the roughness. The fitting is done with piezoelectric actors. These piezoelements can displace the tip with a minimum increment of 10^{-7} mm / V.

- AFM. Basically, this system consists of a cantilever with a tip, a deviation sensor, a piezoactor, and a feedback control (Fig. 4.41). If the tip is within a small distance to the surface, an atomic force develops between the tip and the surface so that the cantilever is bent upwards. A regulator keeps a constant force to the surface of the sample. The input signal for the regulator is laser light, which is reflected by the cantilever and which is sensitive to its position.

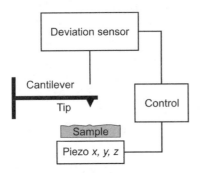

Fig. 4.41   Atomic force microscope (schematic) [74]

Two images of the surfaces of CVD diamond films after thermochemical polishing (Fig. 4.42) are shown as examples of such a measurement. With AFM it is possible to determine a surface roughness in the nanometer range.

4.2.2   Crystallinity

The crystalline state of a material is best investigated with a diffraction experiment (Fig. 4.43). A monochromatic x-ray or electron beam impinging on a crystal with the three primitive axes $\vec{a}$, $\vec{b}$, and $\vec{c}$ is assumed. An electron beam of energy E can be considered as a wave with the wavelength

$$\lambda \ [\text{Å}] = \frac{12}{\sqrt{E \ [\text{eV}]}} .$$ (4.25)

The wave vector of the incoming wave is $\vec{k}$ with $k = 2\pi/\lambda$. The wave vector $\vec{k'}$ of the scattered wave has the same wavelength λ.

The phase difference between the incoming beam serving as a reference and the outgoing beam is

$$\Delta\phi = (\vec{k}\,\vec{a} - \vec{k'}\,\vec{a}) = \Delta\vec{k}\,\vec{a} .$$ (4.26)

Now, all scattered amplitudes from every lattice point must be added together. The vector $\vec{a}$ is generalized to a vector

$$\vec{\rho} = m\,\vec{a} + n\,\vec{b} + o\,\vec{c} .$$ (4.27)

The total amplitude is proportional to

$$\vec{A} = \sum_{\vec{\rho}} e^{-i\,\vec{\rho}\,\Delta\vec{k}} = \sum_{m,n,o} e^{-i\,(m\,\vec{a} + n\,\vec{b} + o\,\vec{c})\,\Delta\vec{k}} =$$

$$\sum_{m} e^{-i\,m\,\vec{a}\,\Delta\vec{k}} \sum_{n} e^{-i\,n\,\vec{b}\,\Delta\vec{k}} \sum_{o} e^{-i\,o\,\vec{c}\,\Delta\vec{k}} .$$ (4.28)

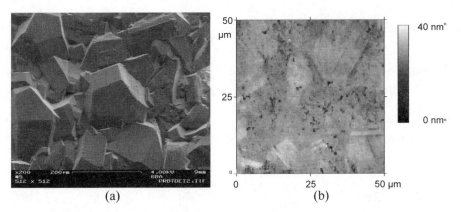

(a)   (b)

Fig. 4.42 (a) SEM images of an as-grown CVD diamond film of optical grade—average surface roughness of 30 μm (profilometer measurement), (b) AFM image of the same surface after thermochemical polishing—average surface roughness of 1.3 nm [75]

The intensity, I, of the scattered beam is proportional to $|A^2|$. Every sum of the right hand side of Eq. 4.28 can be written in the form

$$\sum_{m=0}^{M-1} e^{-im(\vec{a}\,\Delta\vec{k})} = \frac{1-e^{-iM(\vec{a}\,\Delta\vec{k})}}{1-e^{-i(\vec{a}\,\Delta\vec{k})}} \,. \tag{4.29}$$

In order to get the total intensity, we multiply Eq. 4.29 (and the other two sums) by their conjugate complexes. This delivers

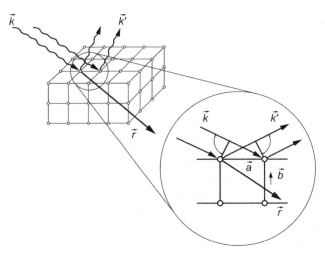

Fig. 4.43   Scattering of a plane wave through a crystal consisting of M^3 atoms

$$I_1 = \frac{\sin^2[M\,(\vec{a}\,\Delta\vec{k})/2]}{\sin^2[(\vec{a}\,\Delta\vec{k})/2]} \,. \tag{4.30}$$

The graph of Eq. 4.30 is a curve with sharp maxima ("lines"). The maxima occur for

$$\vec{a}\,\Delta\vec{k} = 2\pi q \tag{4.31a}$$

where q is an integer. This is one of Laue's equations. The other two are

$$\vec{b}\,\Delta\vec{k} = 2\pi r \;\; \text{and} \tag{4.31b}$$

$$\vec{c}\,\Delta\vec{k} = 2\pi s \,. \tag{4.31c}$$

The curve sketching of Eq. 4.31a shows, for instance, that the maximum height for I_1 is $\propto M^2$, while the width is $\propto 2\pi/M$. I_1 is proportional to its height ($\propto M^2$) times its width ($\propto 1/M$), i.e., proportional to M. Thus, the intensity of the central reflex, I, is proportional to M^3 in three dimensions.

We define a new set of vectors $\vec{A}$, $\vec{B}$, $\vec{C}$, which satisfy the relations

$$\begin{array}{lll} \vec{A}\,\vec{a} = 2\pi & \vec{B}\,\vec{a} = 0 & \vec{C}\,\vec{a} = 0 \\ \vec{A}\,\vec{b} = 0 & \vec{B}\,\vec{b} = 2\pi & \vec{C}\,\vec{b} = 0 \\ \vec{A}\,\vec{c} = 0 & \vec{B}\,\vec{c} = 0 & \vec{C}\,\vec{c} = 2\pi. \end{array} \tag{4.32}$$

If these vectors are additionally normalized in the form

$$\vec{A} = 2\pi \frac{\vec{b}\times\vec{c}}{\vec{a}\,\vec{b}\times\vec{c}}$$

$$\vec{B} = 2\pi \frac{\vec{c}\times\vec{a}}{\vec{a}\,\vec{b}\times\vec{c}}$$

$$\vec{C} = 2\pi \frac{\vec{a}\times\vec{b}}{\vec{a}\,\vec{b}\times\vec{c}}, \tag{4.33}$$

then every vector

$$\Delta\vec{k} = q\,\vec{A} + r\,\vec{B} + s\,\vec{C} \tag{4.34}$$

with integer numbers q, r, and s is a solution of the Laue equations. The vectors $\vec{A}$, $\vec{B}$, $\vec{C}$ define the fundamental vectors of the reciprocal lattice.

It should be noted that only the contribution of the lattice structure to the diffraction pattern has been regarded up to now. Of course, the so-called atomic scattering factors must be considered for the calculation of the expected intensities, i.e., the scattering ability per atom. If two sources of irradiation are compared regarding the investigation of thin films, it turns out that a thin film of a few nanometers diffracts an electron beam so that useful information can be obtained while the same film is not suitable for x-ray diffraction.

X-ray diffraction (XRD). There are several ways to utilize the Laue equations. One of them is *white light* irradiation, i.e., a broad x-ray spectrum is illuminated on the crystal to be examined. The crystal (or rather all sets of lattice planes) interacts with the light of the wavelength (i.e., $\Delta \vec{k}$) that fulfills the Laue equations.

This method offers some advantages such as a quick determination of the crystal symmetry and orientation. As a disadvantage, the lattice constant cannot be determined.

Vice versa, a monochromatic beam can be used, which is exposed on the powder of a crystal to be examined. Therefore, there is always a great number of crystallites (powder grains) in the correct orientation for a given wavelength so that the Laue equations are fulfilled again. The important information is the intensity as a function of the diffraction angle.

For crystalline materials, the wafer is usually rotated (e.g., by an angle θ). Simultaneously, the detector is rotated by an angle 2θ (Bragg-Brentano diffractometer).

An efficient version of XRD is *x-ray topography*. The fundamental idea consists of aligning the crystal which to be examined in such a way that a reflection is measured under a certain angle. A perfect crystal should maintain the diffraction intensity if the beam (or rather the crystal) is shifted laterally. Every imperfection of the crystal violates the diffraction equations. Usually, the structure is created in such a way that a first reference crystal is carefully adjusted so that a sharp monochromatic beam is produced. This one is in turn directed toward the crystal, which should be measured, shifted perpendicularly to the beam. An example is given in Fig. 4.44.

Additionally, the system can be improved by "rocking" the crystal perpendicularly to the plane of incidence (Fig. 4.45). As an advantage of this rocking setup, extremely small deviations in the lattice constant can be picked up. If, for instance, heteroepitaxial films are deposited, two (a doublet line) instead of only one signals are sometimes found. This means that the film still differs from the substrate.

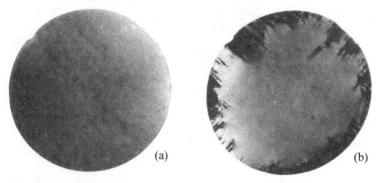

(a)   (b)

Fig. 4.44   (a) Wafer before heat treatment and (b) after formation of dislocations by oxidation at 1200 °C. Examined with x-ray topography [76]

Electron diffraction is performed in vacuum with a monoenergy beam and stationary samples.

In a first version known as low energy electron diffraction (LEED), voltages between 10 V and 1 kV are used. The diffracted electrons are observed in reflection with a fluorescent screen (Fig. 4.46). In order to avoid wrong signals due to surface impurities, the vacuum must be maintained under 10^{-9} Pa. LEED is used in order to measure the so-called surface reconstruction, i.e., the termination of a

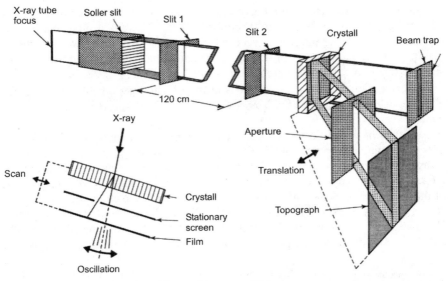

Fig. 4.45   X-ray topography with a rocking setup [77]

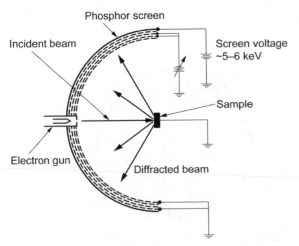

Fig. 4.46   Electron diffraction at low energy (schematic) [78]

surface by a new type of lattice ("superlattice"). A typical LEED pattern is shown in Fig. 4.47.

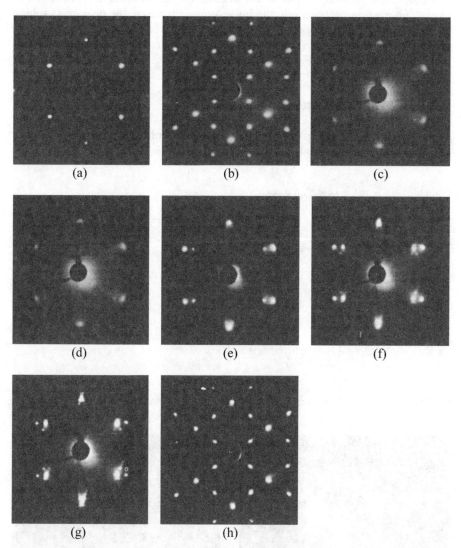

Fig. 4.47 LEED patterns showing the effect of oxygen on the epitaxy of copper on a tungsten (110) surface. (a) Clean tungsten, (b) half a mono layer of oxygen on the tungsten, (c) two mono layers of copper—note the appearance of poorly oriented diffraction spots in the outer copper layer, (d) ten layers of copper, (e) heating to 300 °C for 5 min (resulting in some improvement of the orientation and in the reappearance of the tungsten beams), (f) heating to 550 °C for 15 min, (g) heating to 850 °C for 1 min (tungsten oxide as evidenced by the diagonal rows of beams about the tungsten beam position, (h) heating to 1050 °C for 1 min and returning to half a monolayer of oxygen on the tungsten [79]

Another well-known version is the reflection high-energy electron diffraction (RHEED) as depicted in Fig. 4.48.

Applied energies range from 10 to 100 keV. In order to avoid the penetration of these high-energy electrons into the lower-lying substrates, the operation is done under narrow glancing angles. RHEED is often used for in-situ monitoring of the growth of an epitaxial film (see the section on epitaxy). An example of an RHEED measurement is shown in Fig. 4.49.

4.2.3   Chemical Composition

Secondary ion mass spectroscopy (SIMS) is usually done to detect impurities and their profiles in solid states (Fig. 4.50, [82]).

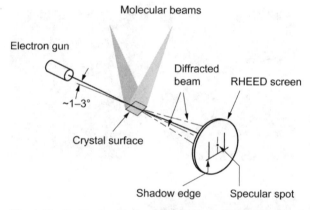

Fig. 4.48   Reflection high-energy electron diffraction (schematic) [80]

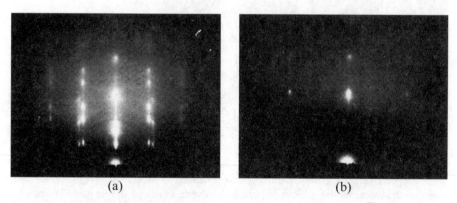

(a)   (b)

Fig. 4.49   RHEED pattern: (00.1) Cu$_2$S overgrowth on (111) Cu. (a) [1$\bar{1}$0] Cu azimuth showing a weak [12.0] pattern of (00.1) Cu$_2$S, (b) [2$\bar{1}$$\bar{1}$] Cu azimuth showing a weak [10.0] pattern of (00.1) Cu$_2$S [81]

Primary ions (PI) from the ion source (IS) are accelerated to the sample (SP) in order to knock atoms off the surface. This process is called sputtering or ion milling. Some of the sputtered ions are electrically accelerated; they can be pulled out to a mass spectrometer (MS) where they are analyzed.

Essentially, SIMS can be operated in two ways: (i) The composition of the wafer (more precisely, that of the surface zone) is determined by the analysis of the secondary ions. This is done with the mass spectrometer (Fig. 4.51). (ii) If the spectrometer is set to depths at a fixed mass, the measured intensity *vs.* sputtering time is a measure of the impurity density *vs.* depth. Calibrated samples are used for the transformation of the ion signal (current from a signal processor) into an ion density. Similarly, the removal rate must be known for the transformation of the sputtering time into a depth.

The great advantage of the method is the possibility to measure all elements (even those in interstitial positions) and the high detection sensitivity of between 10^{15} and 10^{18} cm^{-3}, depending on the element. To obtain a constant atomization rate, it is necessary to use a stable ion gun. The atomization can be achieved either by a finely focused beam, which can be scanned on the surface to be analyzed, or by a beam with a constant current density distribution throughout the radius. In the first case, one has additionally the possibility to record an "ion image" by synchronizing the deflecting plates with an oscilloscope.

The ionization rate strongly depends on the gas coverage on the sample. In particular, oxygen can increase the ionization rate up to a factor of 100. Therefore, sputtering is done with oxygen in some systems. If not, there is always a starting effect always, i.e., an apparently higher foreign atom concentration on the surface due to the unavoidable oxygen allocation.

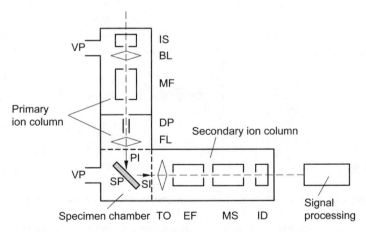

Fig. 4.50 Schematic representation of a SIMS system: PI: primary ions, IS: ion source, BL: beam forming lens, MF: mass filter, DP: deflection plate, FL: focusing lens, SP: sample, SI: secondary ions, TO: transfer optics, EF: energy filter, MS: mass spectrometer, ID: secondary ion detector, VP: vacuum pumps

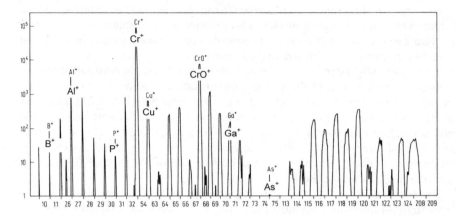

Fig. 4.51   Mass spectrum of a Cr layer of 50 nm thickness on Cu substrate, detected by SIMS [83]

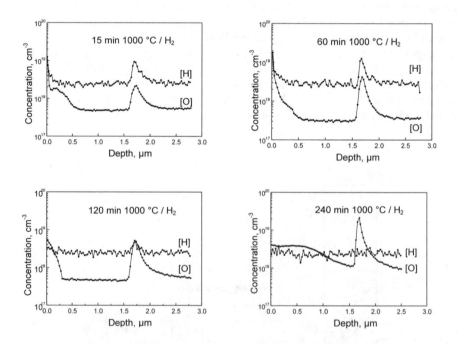

Fig. 4.52   Hydrogen and oxygen profiles after hydrogen implantation in Cz silicon [84]

Noble gases are used as primary ions. The selection of the primary energy is a compromise. On the one hand, it must be large enough for a sufficient sputtering yield. On the other hand, a too high energy simply causes ion implantation and a small sputtering yield. Typical SIMS energies range from 5 to 10 keV. The problems of the procedure are the adjustment of constant sputtering rates, of plane surfaces, and of signals unimpaired by noise. Two examples of the possibilities of SIMS are shown below.

In the first example (Fig. 4.52), an oxygen-rich Cz wafer is implanted with hydrogen and annealed in hydrogen for 15 to 240 min. The depth profiles of water and oxygen, recorded simultaneously, reveals hydrogen to work as a getter center for oxygen. After 120 min the original hydrogen is almost evenly distributed, while the oxygen maintains its sharp profile.

The second example (Fig. 4.53) proves the feasibility of nanolayers by ion implantation. Conventional boron implantation with an energy of 500 eV and plasma-assisted diborane implantation with an energy of 350 eV are done. Activation occurs with rapid thermal annealing in order to prevent the widening of the original Gaussian curve or outdiffusion.

A similar profiling procedure is based on the *Auger electron emission*. Again, sputtering is applied in order to drive through the depth of the layer. However, the signal is now gained by a process that is schematically shown in Fig. 4.54.

An incoming x-ray quantum or a high-energy electron removes an electron from the inner shell (step 1). The deficiency is compensated by an electron from an outer shell (step 2), as shown in the center section of the figure. In a third step, a part of the free energy is transferred to another electron of the outermost shell that gains the remainder as kinetic energy (alternatively, an x-ray quantum can be emitted, right part of Fig. 4.54). The energy in step 2 is the difference of discrete energies, the ionization energy in step 3 is a discrete energy. Therefore, the remaining kinetic energy is likewise a discrete energy. All discrete energies are specific for the respective atoms in whose shell the processes happen. Thus, the kinetic energy identifies the material under investigation like a fingerprint (the same applies to the emitted x-ray quanta). A typical Auger electron spectrum is depicted in Fig. 4.55.

Since the Auger lines are put on a rather broad background, they are detected more easily by differentiating the energy distribution $N(E)$. Therefore, the ordinate of the usual Auger spectrum is converted to the function dN/dE. Electronic dif-

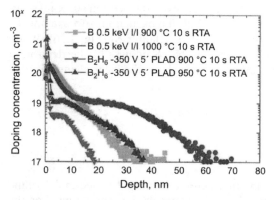

Fig. 4.53   SIMS depth profiles of boron [85]

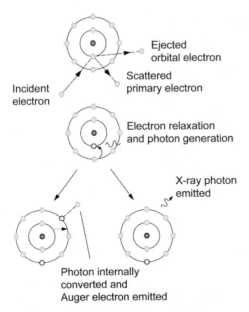

Fig. 4.54   Auger emission process [86]

ferentiation can easily be done with velocity analyzers superimposing a small alternating voltage to the energy-selecting dc voltage and with synchronously recording the output of the electron multiplier. The line height of the Auger signal is usually proportional to the surface concentration of the element causing the Auger electrons.

Contrary to SIMS, Auger electron spectroscopy (AES) can verify the binding state of the material under investigation. The reason is the weak influence of the binding configuration and the neighboring atomic positions on the shell energies. This effect is called *chemical shift.*

Auger measurements can be performed during a sputtering process. The Auger signal of a selected element is measured as a function of the depth. The procedure is similar to SIMS measurement (additionally, the exciting electrons or x-rays can be replaced by sputtering ions. This is called AES ion excitation). An example of such a deep measurement is Ta deposited on polycrystalline Si (Fig. 4.56). When using such a combination, one is interested in the cleanliness of the deposition (i.e., the incorporation of impurity atoms) and in the subsequent chemical reactions between the materials.

4.2.4   Doping Properties

Doping Type

The Seebeck effect, also known as thermoelectric effect, can be used to determine the doping type. Two metal contacts, for example needle tips, are mounted on the

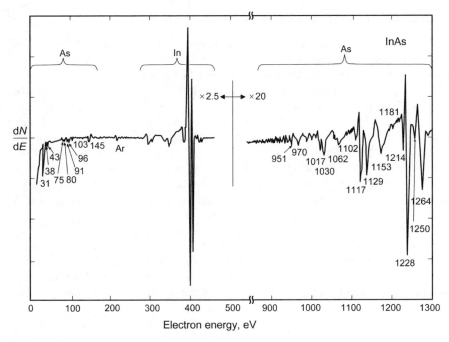

Fig. 4.55   Auger spectrum of InAs [87]

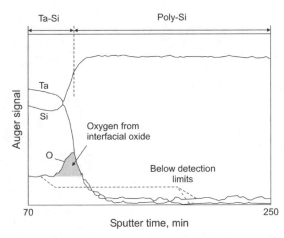

Fig. 4.56   Ta, Si, and O profiles of a TaSi$_2$ film on Si by means of AES [88]

semiconductor. One of the two needles is heated up. With an n-type doping, a positive voltage (and a negative voltage with a p-type doping) is measured at the hot needle relative to the cold needle.

Doping Level, Conductivity, Mobility

In this section, the terms *carrier density* and *doping level* are used without distinction since silicon, the most important material to us, is in saturation at room temperature, i.e., each doping atom contributes one free charge carrier.

Ideally, the semiconductor represents a cuboid with the edge lengths a, b, and c (Fig. 4.57). Two opposite surfaces are brought in contact by evaporation or coating with a conductive paste. The metal must ensure an ohmic contact. No electric rectifying effects or other current-voltage non-linearities may occur. The applied voltage, V, and the subsequent current, I, are measured. The two measured values are converted into the current density j, $j = I/(a\,c)$, and the electric field E, $E = V/b$. From $j = \sigma E$, the conductivity, σ, results. The doping level, N_D or N_A, is obtained from an experimentally acquired reference table (N_D or N_A vs. σ). A direct measurement of the doping level will be shown below.

But technically, a semiconductor is mostly available as a round disk or a rectangular chip with a thickness of some 100 μm. In this case, the four-probe measurement can be conveniently used (Fig. 4.58). Four parallel needles at a distance of 0.635 mm from each other are mounted on the semiconductor. A current, I, is fed through the outer needles resulting in a voltage drop in the semiconductor. The

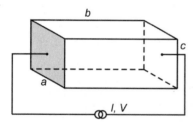

Fig. 4.57   Determination of the conductivity

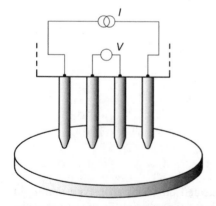

Fig. 4.58   Four-probe measurement (not to scale). If the current feed is also put on the inner needles, a two-probe measurement is performed.

voltage between the inner needles, V, is measured with a high-ohmic voltmeter. The quotient V/I is a direct measure of the specific resistance of the semiconductor, ρ. The correction factor between V/I and ρ must be determined by the potential theory or by comparative measurements. The current and voltage distributions and thus the correction factor depend closely on the distance of the measuring needles and the radius of the disk. For a disk with an infinite radius, the correction factor is 4.53, so that the surface resistance $R_\square = 4.53 V/I$, and the resistivity $\rho = R_\square\, d$ (with a disk thickness d). As an advantage of this procedure, neither the contacts nor any resistances falsify the measurement.

In some cases (e.g., when measuring an inhomogeneous doping process), one has to give up the advantages of the four-probe measurement, and the two-probe measurement is applied. Here, the feed current and the applied voltage are measured in only one pair of points. A correction factor between V/I and ρ is also needed. The pros and cons of the two-point measurement are described in the section on the measurement of impurity concentration profiles.

As stated earlier, the doping values can also be determined from the specific conductivity. In a simplified physical consideration, each doping atom contributes exactly one electron to the increase of the electron density, n, which leads to an increase of the conductivity because

$$\sigma = q\,\mu\,n \tag{4.35}$$

(q is the electron charge, μ the mobility).

The question of which conductivity is obtained with a certain doping (or which measured conductivity corresponds to which doping) can be best answered from experience. The conductivities of several samples of different dopings have been measured, and the curves of ρ vs. N_D and ρ vs. N_A have subsequently been plotted. (for historical reasons, the specific resistance ρ is plotted instead of the conductivity). N_D and N_A are the doping densities for electron and hole doping. These curves are called *Irvin curves*. A representation is depicted in Fig. 4.59 [89].

It should be clear that a current-voltage measurement delivers only the product of the charge carrier density n (or p) and the mobility of the electrons (or holes) when considering Eq. 4.35. To separate the product, a second equation (second measurement) is necessary: the Hall experiment. Basically, let us assume the parallelepiped sample shown in Fig. 4.57. However, a magnetic induction B is applied to the sample in a direction perpendicular to the current flow (Fig. 4.60).

As a result of the Lorentz force

$$F_L = q\,v\,B\,, \tag{4.36}$$

the charge carriers are deflected from their straight-line course between the electrodes. The direction of the Lorentz force is perpendicular to the current and original path (giben by v) and also perpendicular to the field direction *(B)*. The charge carriers are collected on a cuboid surface, whose normal is perpendicular to I (or v) and to B. At the same time, a backward force

$$F_e = q\,E_H \tag{4.37}$$

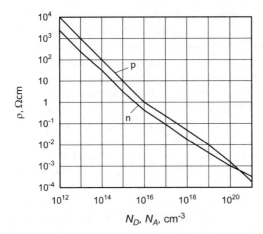

Fig. 4.59 Specific resistance as a function of the doping

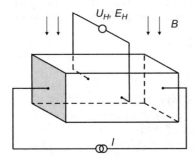

Fig. 4.60 Experimental setup of the Hall effect (schematic)

develops due to an electric field caused by the accumulation of the electrons on the cuboid's surface (this field must not be mistaken with the field causing the current I). In the stationary case, the forces are equal so that $q\,v\,B = q\,E_H$, or after multiplying by the electron density n:

$$(q\,n\,v)\,B = q\,n\,E_H .\tag{4.38}$$

The product $q\,n\,v$ represents the current density. The term R_H determined from the measurements is known as the Hall constant:

$$R_H = \frac{E_H}{j\,B} = \frac{1}{q\,n} .\tag{4.39}$$

The electron density n (the hole density p for a hole semiconductor) is obtained from this term. From the sign of the Hall constant, the information whether electrons or holes are present is obtained.

This procedure can be extended to a simultaneous measurement of the resistivity ρ and the Hall mobility μ_H. The four contacts A, B, C, and D are mounted on

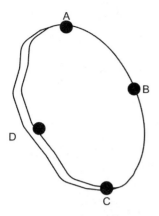

Fig. 4.61   Measurement setup according to van der Pauw [90]

the edge of a thin sample that can have an arbitrary form (Fig. 4.61). The resistance $R_{AB,CD}$ is defined as the quotient of the potential drop $V_D - V_C$ and the feed current between the contacts A and B causing the voltage drop. The resistance $R_{BC,DA}$ is defined analogously. The resistivity

$$\rho = \frac{\pi d}{\ln 2} \frac{R_{AB,CD} + R_{BC,DA}}{2} f \cdot \left(\frac{R_{AB,CD}}{R_{BC,DA}} \right).$$ (4.40)

The factor f is obtained from the implicit equation

$$\cos\left[\frac{(R_{AB,CD} / R_{BC,DA}) - 1}{(R_{AB,CD} / R_{BC,DA}) + 1} \frac{\ln 2}{f} \right] = \tfrac{1}{2} e^{\frac{\ln 2}{f}}.$$ (4.41)

The Hall mobility is determined by measuring the change in the resistance $R_{BD,AC}$ by an applied magnetic field:

$$\mu_H = \frac{d}{B} \frac{\Delta R_{AC,BD}}{\rho}.$$ (4.42)

The method described above is named after van der Pauw [90].

Another way of determining the doping level is the use of an MOS diode. A typical cross section of such a device has already been shown in Fig. 4.27. A bias is applied between the two metal contacts, and an alternating voltage modulation (usually 25 mV) is superimposed on it. In our case, an alternating voltage frequency of 1 MHz is used. The modulation is employed in order to measure the small signal capacitance. This is repeated for all bias values, for instance, between +10 to −10 V, so that a capacitance-voltage curve can be plotted (Fig. 4.62).

The minimum/maximum capacitance ratio C_{min} / C_{max} (also marked with C_{inv} / C_{ox}) is determined by the oxide thickness and, more importantly, by the doping level. With known oxide thickness, this relation can be calculated and pre-

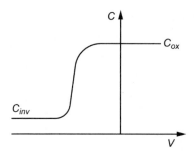

Fig. 4.62   High frequency C-V curve in equilibrium (n-type semiconductor)

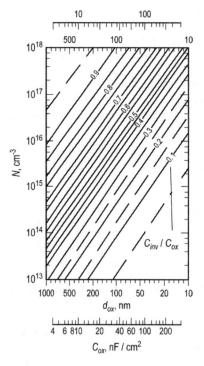

Fig. 4.63   Nomograph for the determination of doping from the ratio C_{min} / C_{max} [91]

sented in the form of a nomograph with the oxide thickness as an additional parameter (Fig. 4.63). The determination of the oxide thickness d_{ox} is discussed in Sect. 4.2.1.

Doping Profile

In some cases, a doping results, whose value changes with the depth into the semiconductor, i.e., a profile (for instance, after diffusion, implantation or oxidation). Important measuring methods are:

(i) *Beveling and two-probe measurement.* In the first step, a probe tip is mounted on the surface of a semiconductor; the lower surface of the semiconductor is grounded (Fig. 4.64). The theory shows that the applied potential of the needle drops about to zero in the semiconductor within a length of approximately three spherical radii of the needle tip. Thus, the resistance between needle and back side is determined only by the resistance within three spherical radii or the doping in this volume. But often there is a high impedance layer, a p-n junction, or an oxide within the semiconductor or on its backside. In this case, the hemispherical potential distribution is violated. Thus, a second needle is usually mounted on the semiconductor surface as counter electrode in a second step. Here, almost half of the applied potential drops symmetrically within three radii (please note the relation to the four-point measurement: the two-points can be compared to the outer needles of the four-point measurement. Thus, the two additional points of the four-point measurement brought in between detect only a small fraction of the potential applied on the outer points).

When performing a profile measurement, that means: the sample, which has been implanted, diffused, etc., is cut into chips and beveled in a third step (Fig. 4.65). The beveling angle depends on the depth profile; typical values are between 0.5° and 10°. The two points are placed on the beveled surface parallel to the beveled edge, and the local resistance is measured. Afterwards, the two needles are moved over the beveled surface and the measurement is repeated. The measurement path along the beveled surface is converted into the depth in the semiconductor (this requires the exact knowledge of the beveling angle). The counting begins from the edge of the beveled surface. In this way, the local resistance is determined as a function of the depth of the semiconductor. The resistance can again be transformed into the doping density using calibration curves. A coverage of the surface with an oxide (as in the case of an MOS structure) is helpful for the determination of the beveled edge since in this case, a clear jump in the resistance occurs. An example of such a measurement (a buried boron layer in n-type silicon) is presented in Fig. 4.65.

(ii) *MOS profile measurement.* The MOS capacitance is the result of a series circuit of the oxide capacitance C_{ox} and the space charge capacitance of the semiconductor C_{sc} (Fig. 4.66).

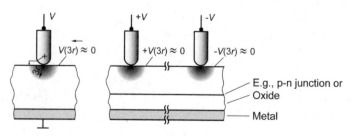

Fig. 4.64 One and two probe measurements

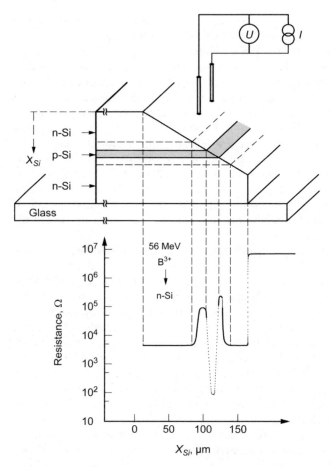

Fig. 4.65 Bevel of an n-type sample implanted with 56 MeV boron and the related resistivity profile. The boron has been activated in nitrogen atmosphere for 30 minutes at 800 °C [92].

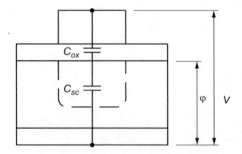

Fig. 4.66 High frequency MOS equivalent circuit (simplified) and voltage definitions

In the following, let us only consider the transition zone of the *C-V* curve, which is rising strictly monotonously (Fig. 4.61), or more precisely, only the section under the so-called flat band point. From these capacitances measured in high frequency C_{hf}, the oxide capacitance is subtracted by applying the series circuit. Thus, the space charge capacitance remains. The latter delivers the position where the doping is measured (where $\varepsilon_{Si} = 1.04 \ 10^{-12} \ \mathrm{F/cm}$):

$$x = \frac{A \, \varepsilon_{Si}}{C_{sc}}.$$

(4.43)

On the contrary, the space charge capacitance can be calculated from Poisson's equation. Let φ be the partial voltage (from the total applied voltage) that drops over the space charge zone in the silicon. The integration of Poisson's equation (first with a constant doping)

$$\frac{\mathrm{d}^2\varphi}{\mathrm{d}x^2} = -\frac{q \, N}{\varepsilon_{Si}}$$

(4.44)

together with Eq. 4.43 delivers:

$$\varphi = -\frac{q \, N \, \varepsilon_{Si}}{2} \frac{1}{C_{sc}^2}.$$

(4.45)

If N is dependent on x and hence on φ, Eq. 4.45 must be written in differential form. The solution for N yields

$$N(x) = \pm \frac{C_{sc}^3}{q \, \varepsilon_{Si}} \frac{1}{\dfrac{\mathrm{d}C_{sc}}{\mathrm{d}\varphi}}.$$

(4.46)

The sign results from the sign of the doping charge. The calculation of the partial voltage φ from the applied voltage V is not trivial. The usual way consists of calculating C_{sc} vs. φ with the doping as a parameter and subsequently C_{hf} vs. V. A slight adjustment for high frequency has to be done, and the free charge carriers must be considered. From the adjustment of the measured values, φ vs. V is obtained. Another procedure circumvents this calculation and the adjustment, but it requires the additional measurement of the low frequency capacity C_{lf} vs. V. Thus,

$$N(x) = \pm \frac{2C_{ox}^2}{q \, \varepsilon_{Si}} \frac{1 - C_{lf}/C_{ox}}{1 - C_{hf}/C_{ox}} \left(\frac{\mathrm{d} \, (C_{ox}/C_{hf})}{\mathrm{d}V} \right)^{-1}$$

(4.47)

and

$$x = \left(\frac{C_{hf}}{\varepsilon_{Si} \, (1 - C_{hf}/C_{ox})} \right)^{-1}.$$

(4.48)

(iii) *Rutherford backscattering (RBS)*. This procedure is shown in Fig. 4.67. First, let us assume an undoped crystal that is exposed to a helium beam (sometimes also a hydrogen beam). The beam is produced by an accelerator. All particles have the same energy E_0 (usually 1 MeV). The sample is aligned in such a way that the particles arrive toward a so-called low-indexed channel axis. In this case, the crystal appears to the particle as consisting of channel tubes and thus, to a large extent, it appears hollow (for a better understanding, please consider a crystal model). Thus, 95 % of the incoming beam can penetrate into the channels, while only 5 % hit the walls of the channels. A small fraction of it is scattered in the direction of a particle detector that is set up to the surface in a direction different from that of the beam.

The detector is energy-dispersive. Thus, it can distinguish the backscattered He particles according to their energy. Since all detected particles are scattered under the same angle and with identical conditions (masses, energies), they have lost the same energy during impact on the surface. Ideally, they all should have the same kinetic energy in the backscattering spectrum. If the near-surface of the crystal is doped, He atoms that already penetrated the channels of the crystal can be scattered again toward the detector on impact with the doping atoms. In particular, this applies to doping atoms that are inserted in interstitial sites, but large—in relation to host lattice atoms—doping atoms inserted in lattice sites also scatter the He beam quite well. On its way in the crystal, the He atom has given off energy. The deeper the scattering position, the smaller the energy that is registered in the detector. From this remaining kinetic energy, the position of the scattering atom can be calculated by the loss of energy per distance. The more doping atoms at this position, the higher the number of particles assigned to this kinetic energy. Thus, the backscattering spectrum is a representation of the profile. It must only be read backwards. Thus, a small energy means a deep position of the doping atoms (Fig. 4.68).

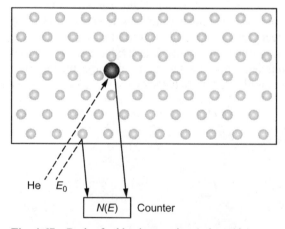

Fig. 4.67 Rutherford backscattering (schematic)

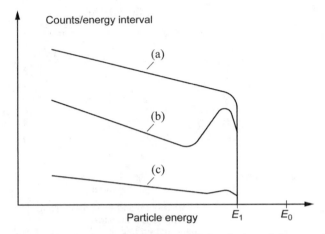

Fig. 4.68 Backscattering spectra: (a) Irradiation of the crystal in a random direction (no channeling), (b) irradiation of an implanted crystal in a channel direction, (c) irradiation of a non-implanted crystal in a channel direction. The energy $E_0 - E_1$ is the recoil energy of a Si surface atom.

The results of an RBS measurement of an Au-Ni-Au layer sequence on a silicon substrate (Fig. 4.69) are shown in Fig. 4.70.

It should be considered how well the Ni layer of 15 nm thickness is detected. As a result of the impact laws, it is apparent that RBS is particularly sensitive to heavy ions.

4.2.5   Optical Properties

Interference colors of thin layers. A first estimate of the thickness of a thin transparent layer deposited on an opaque substrate is obtained based on its color. Therefore, color charts have been developed for the most important cases, SiO_2 and Si_3N_4 [94–96] (Tables 4.3. and 4.4).

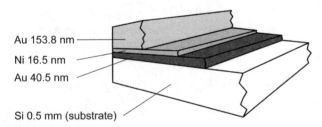

Fig. 4.69   Used layer sequence for an RBS measurement [93]. The lower Au layer of 40.5 nm might rather be labeled with Au-Ge, cf. Fig. 4.70.

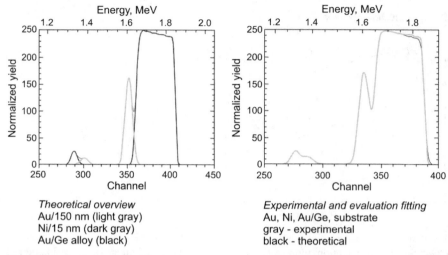

Theoretical overview
Au/150 nm (light gray)
Ni/15 nm (dark gray)
Au/Ge alloy (black)

Experimental and evaluation fitting
Au, Ni, Au/Ge, substrate
gray - experimental
black - theoretical

Fig. 4.70 RBS spectrum for the layer sequence of Fig. 4.69 [93]. The broad curve between the channels 360 and 410 reflects the uppermost Au layer (153.8 nm), the Gaussian-like curve around channel 350 the intermediate Ni layer (16.5 nm). The layer underneath consists of a homogeneous Au/Ge alloy (40.5 nm), it appears as a double peak (channel 290 and 300) since Au and Ge have different recoil energies. The primary energy of the He^{2+} probe beam is 2 MeV.

Table 4.3   Colors of SiO_2

Film thickness, μm	Color	Film thickness, μm	Color
0.05	Tan	0.63	Violet-red
0.07	Brown	0.68	"Bluish" (not blue, but border line between violet and blue green. It appears more like a mixture between violet-red and blue-green and over-all looks grayish.
0.10	Dark violet to red violet	0.72	Blue-green to green (quite broad)
0.12	Royal blue	0.77	"Yellowish"
0.15	Light blue to metallic blue	0.80	Orange (rather broad for orange)
0.17	Metallic to very light yellow-green	0.82	Salmon
0.20	Light gold or yellow-slightly metallic	0.85	Dull light-red-violet
0.22	Gold with slight yellow-orange	0.86	Violet
0.25	Orange to melon colored (dark pink)	0.87	Blue-violet

Table 4.3 (cont'd)   Colors of SiO_2

Film thickness, μm	Color	Film thickness, μm	Color
0.27	Red-violet	0.89	Blue
0.30	Blue to violet-blue	0.92	Blue-green
0.31	Blue	0.95	Dull yellow-green
0.32	Blue to blue-green	0.97	Yellow to "yellowish"
0.34	Light green	0.99	Orange
0.35	Green to yellow-green	1.00	Carnation pink
0.36	Yellow-green	1.02	Violet-red
0.37	Green-yellow	1.05	Red-violet
0.39	Yellow	1.06	Violet
0.41	Light orange	1.07	Blue-violet
0.42	Carnation pink	1.10	Green
0.44	Violet-red	1.11	Yellow-green
0.46	Red-violet	1.12	Green
0.47	Violet	1.18	Violet
0.48	Blue-violet	1.19	Red-violet
0.49	Blue	1.21	Violet-red
0.50	Blue-green	1.24	Carnation pink to Salmon
0.52	Green (broad)	1.25	Orange
0.54	Yellow-green	1.28	"Yellowish"
0.56	Green-yellow	1.32	Sky blue to green-blue
0.57	Yellow to "yellowish" (not yellow but is in a position where yellow is to be expected. At times it appears to be light creamy grey or metallic)	1.40	Orange
0.58	Light-orange or yellow to pink borderline	1.45	Violet
0.60	Carnation pink	1.46	Blue-violet

Table 4.4   Colors of Si_3N_4

Film thickness, μm	Color	Film thickness, μm	Color
0.01	Very light brown	0.095	Light blue
0.077	Average brown	0.105	Very light blue
0.025	Brown	0.115	Light blue-brownish
0.034	Brown-pink	0.125	Light brown-yellow
0.035	Pink-lila	0.135	Very light yellow
0.043	Intensive lila	0.145	Light yellow
0.0525	Intensive dark-blue	0.155	Light to middle yellow
0.06	Dark-blue	0.165	Average yellow
0.069	Average blue	0.175	Intensive yellow

Reflection, absorption, transmission, and refractive index are measured with optical standard devices (e.g., Xe high-pressure lamps, laser, monochromators, spectroscopic ellipsometers). Other systems are developed for Raman, cathodoluminescence, photoluminescence, and electroluminescence measurements. An example of the result of an absorption measurement is given in Fig. 4.71. The steep fall around 1.1 μm reflects the Si gap of 1.12 eV.

Fourier transform infrared spectroscopy (FTIR). FTIR belongs to the paragraph about transmission/reflection measurements. However, it has become so important that it merits special discussion. For this purpose, let us return to the Michelson interferometer of Fig. 4.28. A mirror M_1, which is moved forwards and backwards (variation of L_1) parallel to its perpendicular, is assumed. The mirror M_2 is positioned at a fixed distance L_2. The thickness of the balance disk and the beam splitter is neglected. The sample is held between the beam splitter and detector. For simplification a monochromatic light source of the wavelength λ and the frequency $f = c/\lambda$ is assumed.

Constructive interference results if L_1 equals L_2 (the distance between beam splitters and mirror M_1). If L_1 and L_2, however, differ about a quarter of a wavelength ($\lambda/4$), the path difference is $\lambda/2$, and destructive interference occurs. During the movement of the mirror M_1, an interference pattern of the intensity $I(x)$ is observed behind the sample:

$$I(x) = B(f)\left(1 + \cos\frac{2\pi x\, f}{c}\right). \tag{4.49}$$

$B(f)$ is the product of the light source and the transmission of the sample. The monochromatic light is now replaced with light having a spectral distribution. Then the intensity is modified to

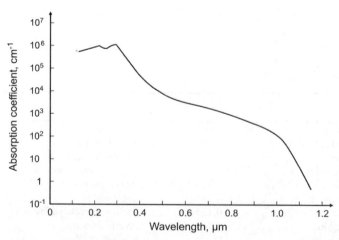

Fig. 4.71   Absorption coefficient of Si at 300 °C *vs.* wavelength [97]

$$I(x) = \int_f B(f') \left(1 + \cos \frac{2\pi x\, f'}{c} \right) df'. \tag{4.50}$$

On the contrary, $B(f)$ is deduced from the measured intensity by reverse transformation.

$$B(f) = \frac{2}{c} \int_{-L}^{L} [I(x') - I_0]\ \cos\left(\frac{2\pi x\, f'}{c} \right)\ dx' \tag{4.51}$$

where

$$I_0 = \int_f B(f')\, df'. \tag{4.52}$$

In order to convert $B(f)$ back to transmission, it must be corrected by the contribution of the light source. This can be done with the measurement of the interference pattern without a sample. At the same time, this procedure eliminates the effects that originate from the atmosphere.

Electron beam induced current (EBIC) and *light beam induced current (LBIC)*. A total surface p-n junction, for instance, a solar cell with a thin emitter, is assumed. The junction is reversed biased for EBIC and short-circuited for LBIC. If an electron beam or a light beam impinges on the junction, the device works as a photodetector (EBIC) or as a solar cell (LBIC). Since the beams are closely bundled compared to the wafer diameter, the measured current can be regarded as a measure of the lateral homogeneity of the junction. When scanning the beam in x and y-directions, an image of the wafer is obtained. Figure 4.72 is an example of

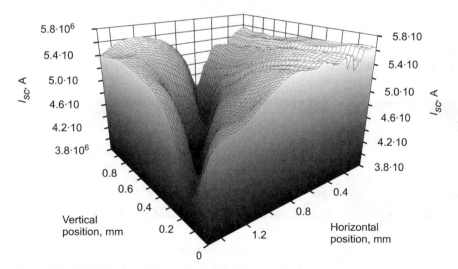

Fig. 4.72 LBIC on a multicrystalline wafer

the image of a multi-crystalline wafer [98]. The grain boundary is clearly visible as a valley.

The Makyoh concept. A magic mirror, which the Japanese called Makyoh, is based on this concept. If one directly looks upon this mirror nothing can be observed. When sunlight is reflected from its front onto a wall, however, the image of a feature appears on that wall. This feature is engraved on the mirror's back (Fig. 4.73).

Visible light is bundled parallelly and directed on the wafer surface to be measured. There, it is reflected and subsequently detected with a CCD camera (Fig. 4.73a). It should be noted that the wafer is held at a certain distance so that the reflected image is outside of the image plane. This shows the latent picture (in our case, a concavity). Concavities appear bright, convexities dark (Fig. 4.73b). Some examples are shown in Fig. 4.74.

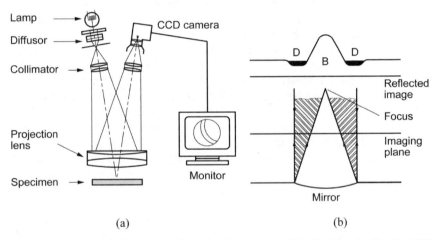

(a)   (b)

Fig. 4.73 Makyoh reflection: (a) Experimental system, (b) principle of reflection [99]

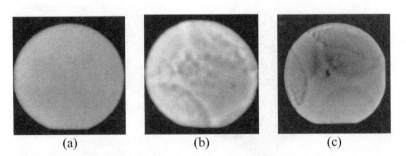

(a)   (b)   (c)

Fig. 4.74 Optical inspection of wafer surfaces. (a) Defect-free wafer, (b) wafer with hills and saw marks, (c) wafer with warpage and local waves after As implantation [99]

4.3   Applications of Nanolayers

Among the three nanostructures nanodefects, nanolayers and nanoparticle, the nanolayers have found the most widespread applications. They are mostly used for electronic or protective purposes:

- MOS gate oxide
- Field oxide
- SOI oxide
- Amorphous layers for heterojunction solar cells, TFTs, optical sensors
- MOS channels
- Counter-doped Si layers for p-n junctions, transistors
- Recrystallized layers on dielectrics for device production
- Oxide as implantation and diffusion masks
- Oxide for the photolithography
- Silicides or metals for connections
- Epitaxial layers for transistors, laser, quantum detectors
- ITO for anti-reflection and charge collection in solar cells
- Back side surface field (BSF) layers in solar cells
- Metal layers for glasses, lenses, beam splitters, interferometers
- Anti-corrosion and passive layers

This enumeration is by no means complete, and each item can be subdivided into numerous applications. For instance, counter-doped layers produced by ion implantation are used for MOS source and drain, CMOS wells, isolation in integrated circuits, buried channel CCDs, layers for the suppression of blooming in the CCD, emitter for solar cells, etc. The number of applications in corrosion and metallization is vast as well. Epitaxial layers enable the production of high frequency, opto-electronic, and quantum layer devices.

4.4   Evaluation and Future Prospects

Historically, the development of thin layers began early and has prospered greatly. Nevertheless, it is afflicted with some problems which will have to be dealt with in future research. In the following, some examples are briefly described:

(i) The scaling and reduction of all geometrical sizes of electronic devices has also lead to the application of thinner gate oxide. It has to be considered, though, that a substantial aspect of gate oxide based MOS technique is based on its high isolation ability. However, when thickness falls short of 4 nm, tunneling currents removing the blocking capability come into action. The only remedy so far consists of accepting these tunneling currents and simply refreshing the signals. So far it is an unsettled question whether more stable dielectrics can be developed.

(ii) The same question refers to the operability of image-delivering CCD devices on the basis of isolation oxides. In this case the signal (the accumulated charge) can flow over the oxide as tunneling current. Thus, apparently the necessary refreshment times are increased but the signal is nonlinearly falsified.

(iii) When dealing with thinner oxides, the above field eventually increases. Effects such as oxide charge trapping and associated bias point shifts can occur then. Current blocking mechanisms such as Coulomb blockade, telegraph noise become apparent among other things.

(iv) Contradictory demands are made concerning the desired size of the dielectric constants of insulators. On the one hand, capacitance values become increasingly small by reducing dimension. Thus, the desire for a high dielectric constant exists. On the other hand, electrical connections are packed more closely together so that the danger of cross talk increases. The capacitive cross-talk is almost proportional to the dielectric constant. Therefore, insulators with small dielectric constant should be used.

(v) In the production of solar cells, antireflection coatings are used as outer layers. A good optical transparency and a small ohmic resistance are required. At present, a transparency of 92 % and a resistivity of $2 \cdot 10^{-4}$ Ωcm are usual. Manufacturers are interested in better values and in a cheaper coating process in order to be able to increase the efficiency and to lower the price of the solar cell.

(vi) A practically unresolved problem still exists in the structuring techniques. A substantial new procedure for this is discussed in Chap. 7. Here, it is only anticipated that it is not sufficient any longer to further develop conventional techniques (lithography, etching). Instead, some new solutions are appearing. An innovative procedure consists of tilting a nanolayer about 90° and using its thickness as a characteristic measure, for instance, for the drain-source separation (spacer technique). Another new procedure places individual atoms next to each other in order to form metallization lines.

(vii) With increasing wafer sizes, inhomogeneities of the layers show to be a problem in signal electronics, power electronics, and also in solar cells. At the same time, problems occur in the detection technique. During deposition, for instance, a mapping representation, i.e. an image of the topography of a parameter of the deposited layer, is frequently required. With a lateral resolution of 10 µm, this can be extremely time-consuming in the case of a 12″ wafer.

(viii) With decreasing thickness of a series of thin layers manufactured by epitaxy or CVD, diffusion becomes a serious problem.

(ix) A limiting factor in the efficiency of a processor is its communication with other parts of the system. According to Rent's rule, the number of external connections N_i should reach at least

$$N_i = 2.5 N_g^{0.6} \qquad\qquad (4.53)$$

whereby N_g is the number of gates on the processor. For today's chips, this number is not achieved by about an order of magnitude. Even with further progress in the structuring of metal layers, no improvement can be expected. On the contrary, the situation is worsened by the fact that with each reduction of the structures, the number of gates increases quadratically (with the surface) but the number of connections do so only linearly (with the edge length).

A promising way out of this dilemma can be the application of free space optics. The electrical signal on a processor is converted into an optical signal and directed on a second chip facing the processor. From there, it is directed to the desired position in the processor by a lens system [100, 101]. Further advantages with this system are higher bandwidths and the reduction of power losses.

5 Nanoparticles

5.1 Fabrication of Nanoparticles

At first sight a nanoparticle is defined as a ball or a ball-like molecule which consists of a few 10 to some 10,000 atoms interconnected by interatomic forces but with little or no relationship to a solid state. However, this intuitive concept is not fulfilled in many cases. A first example is a nanocrystalline Si particle which is embedded into an amorphous matrix. Other examples are nanoparticles which are compressed to bulk ceramic or surface layers. It should be considered that deposited nanoparticle layers differ from uniform layers, particularly due to the presence of grain boundaries, which leads to different electrical and optical behavior.

5.1.1 Grinding with Iron Balls

First of all, a container is filled with stainless steel balls of a few millimeters in diameter. The material to be crushed is added in the form of a powder of about 50 μm diameter grain size. After filling the container with liquid nitrogen, a rotating shaft grinds the material. The grinding periods are within the range of minutes to some 100 hours. This process is simple; its weakness, however, lies in the fact that the grinding balls contribute to impurities.

5.1.2 Gas Condensation

A typical system is shown in Fig. 5.2. The operation occurs in an evacuated chamber with a pressure of 10^{-5} Pa. After mounting the raw material on one or more crucibles, it is evaporated thermally, by an electron gun, or by ion sputtering. The evaporated atoms or molecules unite and form particles of different sizes. Finally they are captured with a cold finger from which they are scraped off and collected with a funnel. The particle diameter is usually within the range of 5 to 15 nm.

5.1.3 Laser Ablation

The raw material is provided as a solid. Its dissolution is achieved by a focused laser beam—similar to the cutting of a metal or a semiconductor. The advantage

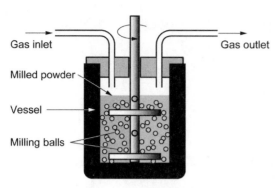

Fig. 5.1   Ball mill for the fabrication of nanoparticles [102]

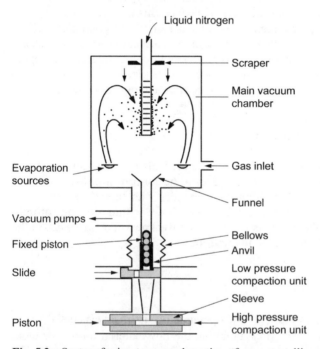

Fig. 5.2   System for inert gas condensation of nanocrystalline powder [103]

of this procedure is a 1:1 transfer of the material composition from the raw material to the particles. The system has already been shown in Fig. 4.8.

5.1.4   Thermal and Ultrasonic Decomposition

As an example of thermal decomposition, the starting material iron pentacarbonyl, $Fe(CO)_5$, is considered. It decomposes in a polymeric solution, e.g., polybutadiene, and thus produces iron particles of 7 to 8 nm diameter. The material can also be decomposed as immersion in decane by ultrasonic irradiation, resulting in particle sizes roughly between 6 and 240 nm. Smaller values apply to the more highly solved systems.

5.1.5   Reduction Methods

Some metal compounds (e.g., chlorides) can be reduced to elementary metallic nanopowder by the application of $NaBEt_3H$, $LiBEt_3H$, and $NaBH_4$, for example. The reaction equation can be written as (M: metal, Et: ethyl):

$$MCl_x + x\,NaBEt_3H \rightarrow M + x\,NaCl + x\,BEt_3 + \tfrac{1}{2}x\,H_2 \qquad (5.1)$$

5.1.6   Self-Assembly

This phenomenon is usually found in the heteroepitaxy. Three-dimensional islands are formed with a rather surprising regularity on a substrate. Due to the free surface energies (substrate–vacuum, substrate–film, film–vacuum), two extremes can occur: regular layer-on-layer growth and cluster formation. The first case is comparable with the picture of butter on bread, and the second one with water drops on butter. In the case of a large lattice mismatch, an intermediate case can occur. The film follows a layer-on-layer growth but develops greater and greater pressure. With sufficient film thickness the film will form three-dimensional islands as in the preceding second case (Stranski-Krastanov mechanism, Fig. 5.3). An example of these self-assembly islands is presented in Fig. 5.4.

5.1.7   Low-Pressure, Low-Temperature Plasma

Although plasma excitation can take place with direct or alternating current, normally a conventional capacitively coupled RF plasma device is used. A gas (e.g., silane) is let in so that a pressure of 1 to 200 Pa develops. As shown in Sect. 4.1.1, the gas is ionized by the applied field. The free electrons acquire kinetic energy

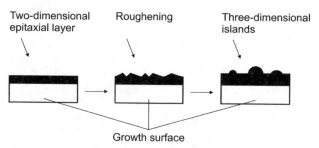

Fig. 5.3 Stranski-Krastanov growth [104]

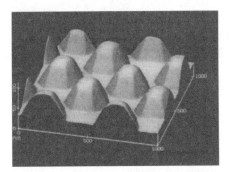

Fig. 5.4   Self-assembled $As_{0.5}Ga_{0.5}As$ islands on $In_{0.2}Ga_{0.8}As$ [105]

and, in return, ionize neutral molecules upon impact. Due to their mass, the ions are slow and maintain the temperature of the gas. This justifies the name of low-temperature plasma.

In the applications described in Sect. 4.1.1, interest is focused on the deposited layers. Here, however, we are dealing with the ionized fragments of the gas which maintains the plasma. As an example, the fragments from silane after electron collision are listed in Table 5.1. These ions and their agglomerates form the nanoparticles which can then be examined.

The particles produced are accumulated mainly close to the plasma layer of the power-operated electrode (this is a sign that the fragments are negatively charged), where they can be measured mostly in situ. After extinguishing the plasma, however, it is possible to collect some particles which had fallen onto the substrate.

5.1.8   Thermal High-Speed Spraying of Oxygen/Powder/Fuel

Thermal spraying is a procedure for nanocrystalline cover layers. A burn reaction produces high temperature and high pressure within a spray gun. The pressure

Table 5.1   Dissociation products after impact of an electron with silane [106]

Products	Threshold energy, eV	
$SiH_2 + 2H + e^-$	8	(?)
$SiH_3 + H + e^-$	(?)	
$SiH + H_2 + H + e^-$	10	(?)
$Si + 2H_2 + e^-$	12	(?)
$SiH^* + H_2 + H + e^-$	10.5	
$Si^* + 2H_2 + e^-$	11.5	
$SiH_2^+ + H_2 + 2e^-$	11.9	
$SiH_3^+ + H + 2e^-$	12.3	
$Si^+ + 2H_2 + 2e^-$	13.6	
$SiH^+ + H_2 + H + 2e^-$	15.3	
$SiH_3^- + H$	6.7	
$SiH_2^- + H_2$	7.7	

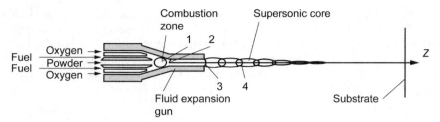

Fig. 5.5 Spray gun for thermal high-speed spraying of oxygen/powder/fuel [107]

drives particles of a nanopowder contained in the pistol through a nozzle onto the surface which is to be covered. An example of a spray gun is shown in Fig. 5.5.

5.1.9 Atom Optics

An atom beam which is produced by heating up a material in a crucible is assumed. The beam is bundled by one or more apertures and steered onto the substrate (Fig. 5.6). In the next step, the beam is subjected to the dipole forces of a standing wave which is produced by a laser beam (Fig. 5.7). The atoms will devi-

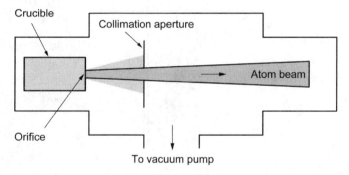

Fig. 5.6 Manufacture of an atom beam [108]

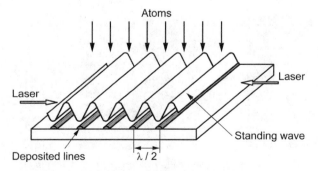

Fig. 5.7 Lens arrangement with a standing wave [108]

ate to the nodes of the standing wave where they are subjected to the smallest forces. The ideas of Fig. 5.6 and 5.7 can be combined for the assembly of atomic bundling (Fig. 5.8). A lattice pattern reflecting the wavelength of the laser is produced (Fig. 5.9). A second standing wave may be used perpendicularly to the first and to the direction of the beam. This gives rise to a two-dimensional pattern of deposited atoms (Fig. 5.10).

5.1.10   Sol Gels

A sol (hydrosol) is a colloidal dispersion in liquid. A gel is a jelly-like substance formed by coagulation of a sol into a gel.

The best known example of a sol gel process is probably the production of SiO_2. A catalyst (acid or base) is added to a solution of tetramethoxysilane (TMOS), water, and methanol. Hydrolysis of the Si–OMe (Me: methyl) bonds

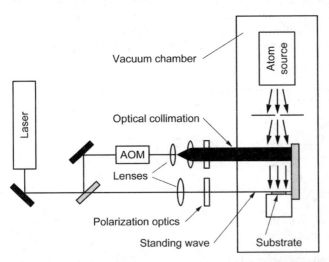

Fig. 5.8   Assembly for atom focusing in a standing wave [108]

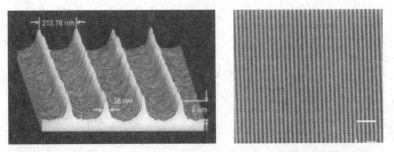

Fig. 5.9   AFM images of a one-dimensional lattice with 212 nm pitch and 38 nm line width formed by laser-focused atom deposition of chromium [109]

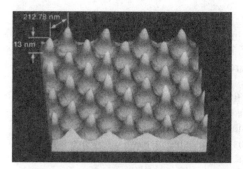

Fig. 5.10   AFM images of a two-dimensional lattice formed by laser-focused Cr deposition [110]

leads to the formation of Si–OH groups:

$$Si(OMe)_4 + 4H_2O \rightarrow \text{"}Si(OH)_4\text{"} + 4MeOH \tag{5.2}$$

Further dehydration reduces the "$Si(OH)_4$" to SiO_2 gel. If hydrolysis and condensation are completed, a silicon oxide xerogel is formed (in Greek, *xeros* means dry). During the reaction, the gel reaches a viscosity so low that it can be applied onto a centrifuge and distributed over the wafer. When annealing over 800 °C, homogeneous oxides, comparable to bad MOS gates, can be manufactured. If doped SiO_2 layers are produced, they can be used as diffusion sources in a subsequent process.

At moderate and low solidification temperatures, the procedure delivers the so-called *nanocomposite*. By definition, nanocomposites contain nanoparticles of less than 1000 nm in a host matrix. The following nanocomposites have already been manufactured (the list is not complete) [111]: nano-Co/Mo, Cu, Fe, Ni, Pd, Pt and Ru in Al_2O_3, SiO_2, TiO_2 and ZrO_2 gels, nano-C, Cu/Ni, Pd/Ni, and Pt in silica gel, nano-Ag, Ge, Os, C, Fe, Mo, Pd, Pt, Re, Ru and PtSn in silica gel-xerogel.

5.1.11 Precipitation of Quantum Dots

Quantum dots are three-dimensional semiconductor materials in or on a matrix. Nanocomposites from semiconductor materials and the above-mentioned self-assembled islands belong to this group. Sometimes it is difficult to differentiate between quantum dots and nanodefects. An example is SiO_2 implanted with Ge, which is used for photoluminescence experiments (Sect. 3.3.4).

The earliest descriptions of quantum dots took place with the investigation of semiconductor precipitation in glasses. Precipitation is still used in the manufacturing of CdS, CdSe, CdTe, GaAs, and Si nanocrystallites in silica glasses. The contaminants are added to the melt, and after a further annealing step from 600 to 1400 °C, they form precipitates of controllable size, for instance, 2 nm for CdTe dots in boron silicate glass.

Another procedure is codeposition of quantum dots with thin films. There are several modifications, but the common principle is the production of nanocrystallites in a separate step (by evaporating, laser ablation, sputtering a target, etc.). They are directed towards a substrate which is, however, covered at the same time with a film so that they are included into this film.

Quantum dots can also be manufactured by means of lithography. Since a high resolution is required, electron beam lithography must be employed. The procedure takes place mostly in such a way that the material which is to be converted into quantum dots is deposited on a substrate like GaAs via MBE. The size of the electron beam spot determines the smallest possible size of the quantum dot. Therefore, if the wafer is etched, islands of this size remain. They are further reduced by etching so that quantum dots of a few 10 nm can be produced. Very often, the layers are covered in order to improve the results.

5.1.12 Other Procedures

GaN nanoparticles were synthesized in autoclaves with the reaction of Li_3N and $GaCl_3$ in benzene at 280 °C. The particle size was approximately 32 nm [112]. The same material was obtained by pyrolysis of polymeric galliumimide, {Ga $(NH_3)_{3/2}\}_m$ in the presence of NH_3 [113].

Fullerenes (Sect. 2.2) are acquired by arc discharge in plasma or by an atomic beam furnace (Fig. 5.6) filled with carbon. After dissolution of the soot particles in an organic liquid, the fullerenes can be separated by gas chromatography.

5.2   Characterization of Nanoparticles

5.2.1   Optical Measurements

Nanoparticles are more or less characterized in the same way as nanodefects or nanolayers. In their case, size effects are more apparent. Therefore, in view of Sect. 2.1, they are attractive, namely, to pursue the transition from the solid state to the nano-behavior.

A first comparison is obtained from IR measurements of an a-Si:H film and silicon nanopowder (Fig. 5.11) [106]. Three bands appear, namely stretching bands, bending bands, and wagging bands. These bands can be assigned to specific groups of hydride and polymer chains, and their presence reveals information on the specific nanostructure. Please note the band between 840 and 910 nm [a proof of SiH_2 and $(SiH_2)_n$], which is missing in a-Si:H. A similar comparison can be drawn from the luminescence spectra of bulk and nanocrystalline Si and the absorption spectra of bulk and nanocrystalline Ge (Fig. 5.12).

Optical absorption measurements provide additional information about the development of the solid state parameters, depending on cluster size. An example is the measurement presented in Fig. 5.13 where optical density of CdSe *vs.* light energy is plotted [116]. The blue shift with reducing cluster size is a generally recognized phenomenon. However, we kindly refer the reader to the measurements in Sect. 2.3 [9] where the opposite behavior is described.

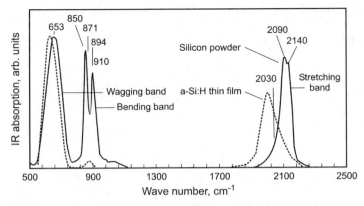

Fig. 5.11   IR spectra of silicon nanopowder in comparison to a standard a-Si:H film

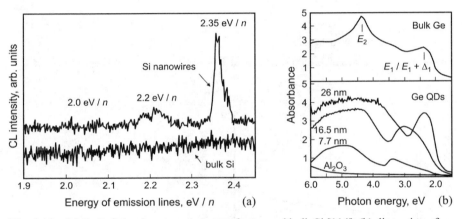

Fig. 5.12   (a) Cathodoluminescence spectra of nano and bulk Si [114], (b) dispersion of bulk Ge and absorption of nano Ge in an Al_2O_3 matrix [115]

5.2.2   Magnetic Measurements

The measurement of the ferromagnetism of nanosize iron and the comparison with bulk iron is shown in Fig. 5.14 [117].

It is clearly seen that the saturation behavior is lost with decreasing grain size. The authors explain this behavior as the transition from ferromagnetism to superparamagnetism.

5.2.3   Electrical Measurements

Only one example is shown here since electrical devices form a subsequent chapter of this work.

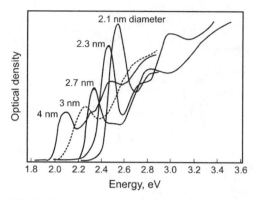

Fig. 5.13   Absorption spectrum of CdSe for numerous sizes of nanoclusters.
The measurements are performed at 10 K.

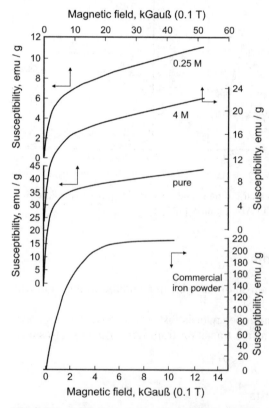

Fig. 5.14   Saturation curves of nanosize iron and macroscopic iron powders. M is for the
molecular concentration of the raw $Fe(CO)_5$ solution in decane; the particle size decreases
with reducing molecular fraction.

Due to the reduction of dimensions, the particles are subjected to a transformation of the band structure as shown in Sect. 2.3. This applies particularly to metal particles whose electrons can be thought to be locked up in a box with high walls where standing waves develop as eigenfunctions. Such a system can controllably exchange single electrons with a second one by tunneling through the walls. The physics of current transport will be discussed later. However, the measurement setup and the arising I-V curves of the so-called ligand-stabilized Au_{55} cluster (the ligand is a colloidal chemical stabilizer) are already depicted in Figs. 5.15 and 5.16, respectively.

5.3   Applications of Nanoparticles

Nanoparticles are already applied as:

- Optical filters in sunscreen and skin cream

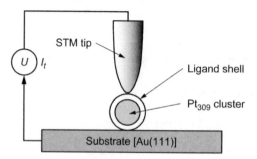

Fig. 5.15   STM as a tool for the measurement of single electrons at a ligand-stabilized Pt_{309} nanocluster [118]

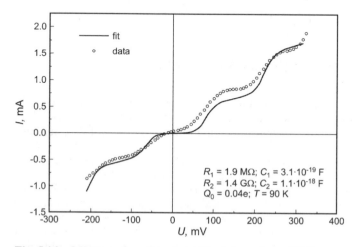

$R_1 = 1.9\ M\Omega;\ C_1 = 3.1 \cdot 10^{-19}\ F$
$R_2 = 1.4\ G\Omega;\ C_2 = 1.1 \cdot 10^{-18}\ F$
$Q_0 = 0.04e;\ T = 90\ K$

Fig. 5.16 I-V curves for a ligand-stabilized Au_{55} cluster [118]

- Dirt repellents for cars and windows, Fig. 5.17 [119]
- Flat screens, Fig. 5.18 [120]
- Single electron transistors

Research is underway in the area of

- Nanowheels, nanogears, nanofilters
- Drug pumps located in the human body, long-term depots
- Luminescent devices (after installation in zeolites)
- Energy storage (hydrogen in zeolites)
- Electronic devices

The following applications are expected particularly for the fullerenes:

- Particle absorption filters for cigarettes
- Chromatography
- Molecular containers
- Sensor cover layers for surface wave devices
- Additives in fuels
- Lubricants
- Catalysts for hydrogenation
- Photocatalysts for the production of atomic oxygen in laser therapy
- Production of artificial diamonds
- Functional polymers, photoconductive films
- Alkali metal MC_{60} chain formation (linear conductivity)
- Superconductivity (doping with alkali metals)
- Ion engines
- Raw material for AIDS drugs
- Tools that are harder than diamond
- Nanoelectronic devices

Again, the enumeration is incomplete.

5.4   Evaluation and Future Prospects

In order to evaluate the progress in the area of nanoparticles, we quote literally from a publication on the nanoparticle industry's total revenue [106]. Based on these figures, we may draw a conclusion on the future development with due caution. The quotation reads:

"According to the technical-market study 'Opportunities in Nanostructured materials', published recently by the Business Communications Co, the overall U.S. market for nanostructured particles and coatings was valued at an estimated 42.3 million US$ for 1996. This tabulation included ceramic, metallic, semiconducting, and diamond nanostructured materials produced in commercial quantities with the exception of nanoscale amorphous silica powder, which commands a market of several hundred million dollars. The market for nanostructured materials is projected to grow about 400 % in 5 years. It is expected to

Fig. 5.17   Partial coverage of a car door with a layer of nanoparticles for dirt rejection

Fig. 5.18   Flat screen from nanotubes

Table 5.2   Overall U.S. markets for nanostructured materials

	1996 US$ (Mio)	%	2001 US$ (Mio)	%	Average annual growth rate, %
Particles[a]	41.3	97.6	148.6	96.1	29.2
Coatings	1.0	2.4	6.0	3.9	43.1
Total	42.3	100	154.6	100	29.6

[a] Dry powders and liquid dispersion

reach $154.6 million in 2001, corresponding to an average annual growth rate of 29.6 % from 1996 to 2001."

Table 5.2 sums the overall U.S. markets for nanostructured materials. It should be noted that coatings obviously represent layers which contain nanoparticles (an example would be vapor-deposited fullerenes). Moreover, *nanostructured materials* can be equated with nanoparticles. According to unpublished sources, these figures were exceeded by far.

6 Selected Solid States with Nanocrystalline Structures

6.1 Nanocrystalline Silicon

6.1.1 Production of Nanocrystalline Silicon

Silicon with nanoscale crystalline grains—ranging in size from a few nanometers to 1000 nm—is referred to as nanocrystalline silicon (nano Si, n-Si). This form of silicon is usually composed of deposited layers of approximately 1 μm thickness each. On the contrary, when exceeding 1000 nm in grain size, we enter the domain of microcrystalline silicon. In this paragraph, production and analysis of nanocrystalline silicon are discussed, essentially based on [121]. It can be manufactured using different deposition methods such as

- Electron cyclotron resonance CVD (ECRCVD) [122],
- Photo CVD [123],
- Magnetron plasma CVD [124],
- Plasma-enhanced CVD (PECVD) [125, 126],
- Remote plasma-enhanced CVD [127],
- Hydrogen radical CVD [128],
- Spontaneous CVD [129], and
- Reactive sputtering [130, 131].

In the following, emphasis is put on the PECVD procedure (cf. Sect. 5.4), which is done very similarly to the deposition of amorphous silicon. As a major difference, however, higher frequencies (e.g., 110 MHz) are preferred during nanocrystalline deposition instead of the usual 13.56 MHz. While being helpful for the production of films, this measure causes a problem regarding homogeneity: with 13.56 MHz the wavelength is 22 m, and it shortens to 2.7 m with 110 MHz. Thus, it lies within the limits of the chamber size and standing waves can develop. Consequently, precautions must be taken for large-surface deposition with frequencies above 60 MHz. For instance, by a multi-point feed of the HF power with same amplitude and phase, the homogeneity of the HF potential and hence the deposition process can be improved.

The raw material for the production of silicon layers is usually silane (SiH_4). For the deposition of doped layers, phosphine (PH_3) for n-type layers and diborane (B_2H_6) or trimethylborane [$B(CH_3)_3$] for p-type layers is added to the silane. Silane is a pyrolytic gas which reacts explosively with air or water vapor.

The massive dilution of the silane with hydrogen makes the second major difference. Contrary to argon or helium, this is a reactive dilution gas in the plasma which is able to react and intercept with radicals being detrimental for the layer deposition. Hydrogen can also restrict deposition and etch silicon.

Investigations of hydrogen-diluted silane in connection with the deposition of a-Si:H have already been undertaken earlier by Chaudhuri *et al.* [132] and Shirafuji *et al.* [133]. There, the aspect of the degree of hydrogen dilution was not considered sufficiently.

With a dilution $R = [SiH_4]/[H_2]$ of >0.05, a-Si:H is formed using PECVD deposition, and this stability field with reference to the structural and electrical characteristics can be further subdivided. The positive effect on a-Si:H is the reduction of $(SiH_2)_n$ chains. Furthermore, the density of the crystalline silicon increases up to 90 %, the refractive index increases, and the total hydrogen quantity built up in the structure decreases [134].

A strong dilution of the starting gas, silane, during plasma deposition can modify the structure of the deposited film from a pure amorphous to a mixture of amorphous and crystalline phases. Generally, nanocrystalline silicon is deposited under strong dilution of the silane with hydrogen (ratio 1:100–10,000) and under depletion of silane [135]. In the VHF range, nanocrystalline growth for a silane dilution of <7.5 % has already been observed [136]. At standard frequency, Tsai *et al.* [137] found that nanocrystalline growth begins at silane dilutions of ≤4 %.

The exact mechanism of how hydrogen improves the material properties in the silicon structure is still being debated. However, it is generally assumed that amorphous and micro to nanocrystalline silicon are deposited simultaneously. Strong dilution and hydrogen reduce the amorphous fraction and can even lead to its etching.

On the one hand, hydrogen dilution leads to a reduction of the deposition rate, while on the other hand, it can prevent inhomogeneities of the silane and, thus, the formation of particles in the gaseous phase is reduced. This proves favorable, especially for the deposition of large areas [138].

In the following, some data about the thicknesses and deposition rates for different variations of deposition conditions are shown in Tables 6.1–6.6 [121]. The excitation frequency is 110 MHz (with the exception of frequency variation), coupled plasma power 10 W, power based on electrode surface 0.7 W/cm^2.

6.1.2   Characterization of Nanocrystalline Silicon

The measurement methods performed on monocrystalline wafers can be transferred to nanocrystalline layers. Some examples are shown below.

Diffraction. A nanocrystalline layer is deposited on a Dow Corning glass 7059 with 110 MHz and subsequently examined with x-rays [139]. The results are depicted in Fig. 6.1.

An amorphous background and a superimposed Bragg reflex usually show up. From this measurement, it is not possible to know to what extent the amorphous dispersion of the glass contributes to the entire amorphous signal. As can be ex-

Table 6.1 Thickness and deposition rate of n-Si (n-type) by variation of the deposition pressure. Temperature 280 °C, silane fraction in hydrogen 1.93 %, PH$_3$-fraction in silane 1.5 %

p_{dep}, mTorr	Thickness, nm	Rate, pm/s
250	309	172
300	298	166
350	275	153
400	295	164
450	212	118
500	208	116
550	inhom.	-
600	202	112

Table 6.2 Thickness and deposition rate of n-Si by variation of deposition temperature. Pressure 450 mTorr, silane fraction in hydrogen 1.93 %, PH$_3$-fraction in silane 1.5 %

T_{dep}, °C	Thickness, nm	Rate, pm/s
280	301	167
300	310	172
320	311	173

Table 6.3 Thickness and deposition rate of n-Si by variation of the PH$_3$-fraction in silane. Temperature 280 °C, pressure 350 mTorr, silane fraction in hydrogen 1.94 %

Gas flow, sccm SiH$_4$:H$_2$:PH$_3$[a]	Doping $\dfrac{[PH_3]}{[PH_3]+[SiH_4]}$, %	Thickness, nm	Rate, pm/s
3:200:1	0.75	324	180
2:150:1	1.0	312	173
1:100:1	1.5	279	155
1:124:1.5	1.8	214	119
1:149:2	2.0	313	174
1:197:3	2.25	321	178
0.5:123:2	2.4	300	167
0.5:148:2.5	2.5	295	164
147:3	3.0	255	142

[a] PH$_3$: 3 % PH$_3$ in SiH$_4$

Table 6.4 Thickness and deposition rate of n-Si by variation of the excitation frequency. Temperature 320 °C, deposition pressure 350 mTorr, silane fraction in hydrogen 1.94 %, PH$_3$-fraction in silane 2.5 %

Frequency, MHz	Thickness, nm	Rate, pm/s
110	376	157
90	329	137
70	247	103
50	269	112

Table 6.5 Thickness and deposition rate of n-Si by variation of the silane fraction. Temperature 320 °C, deposition pressure 350 mTorr, PH_3-fraction in silane 3 %

Series	Gas flow, sccm H_2:PH_3	Silane fraction $\dfrac{[SiH_4]}{[SiH_4]+[H_2]}$, %	Thickness, nm	Rate, pm/s
	200:2	0.96	112	62
Dilution	200:3	1.43	232	129
	200:4	1.90	199	111

Table 6.6 Thickness and deposition rate of n-Si by variation of the deposition temperature duration. Temperature 280 °C, deposition pressure 400 mTorr, silane fraction in hydrogen 1.93 %, PH_3-fraction in silane 1.5 %

Deposition time, min	Thickness, nm	Rate, pm/s
10	79	132
20	144	120
30	218	121
40	296	123
60	438	122

Table 6.7 Thickness and deposition rate of n-Si (p-type) by variation of temperature. Gas flow SiH_4:H_2:$B(CH_3)_3$ 2:200:2 sccm [$B(CH_3)_3$: 2 % in He], deposition pressure 200 mTorr, silane fraction in hydrogen 0.99 %, B_2H_6-fraction in silane 1.96 %

T_{dep}, °C	Thickness, nm	Rate, pm/s
400	464	130
380	490	140
360	453	130
340	479	130
320	555	150
300	537	150
280	337	90
260	406	112
240	430	120
220	449	120

pected, only planes with even or odd numbers contribute to the reflection because of the fcc structure of the silicon.

Apart from the determination of the crystal structure, diffraction provides yet more information. The crystallite sizes δ_{hkl} are determined using the Debye-Scherrer formula [140] by the equation

$$\beta_{hkl}(2\vartheta) = \frac{\kappa\,\lambda}{\delta_{hkl}\,\cos\vartheta} \qquad (6.1)$$

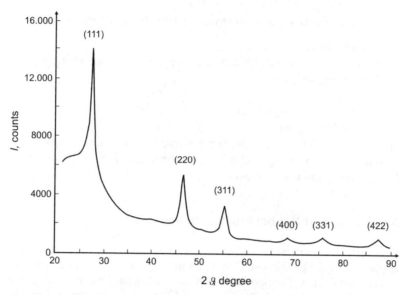

Fig. 6.1   X-ray diffraction of an n-Si layer deposited on a Dow Corning glass

(where κ is the form factor and has a value ranging from approximately 0.89 to 1.39, λ the wavelength, θ the diffraction angle, h, l, and k the Miller indices, and β the half width of the respective x-ray peaks). They vary depending upon deposition frequency by (6.7 ± 0.5) nm at 50 MHz, (7.5 ± 0.5) nm at 70 MHz, (7.3 ± 0.4) nm at 90 MHz, and (7.5 ± 1) nm at 110 MHz.

Further information concerns the ratio of crystalline to amorphous fractions. As already described above, the amorphous fraction cannot be clearly described. However, an estimation is obtained if the peaks' ratio (e.g., 220 to 111) is compared. The ratio of the crystalline to amorphous intensity (I_{cr}, I_{am}) is obtained from Table 6.8.

The samples deposited with low deposition pressure manifest twice as large a fraction of crystalline silicon as the samples deposited at high deposition pressure. For doped samples with the lowest phosphorus content, the amorphous fraction is only weakly pronounced and can no longer be evaluated with the above-mentioned method. No crystal oriented anisotropy can be detected.

Nanocrystalline silicon layers have been and still are the subject of numerous investigations. An in-depth discussion is beyond the scope of this book; however,

Table 6.8   Ratio of crystalline to amorphous scattering intensity

Variable parameter	280 °C	320 °C		250 mTorr	600 mTorr
I_{cr}/I_{am}	1.15	1.6		2.25	0.9

Variable parameter	PH$_3$ in silane 3.0 %	90 MHz	70 MHz	50 MHz
I_{cr}/I_{am}	1.9	2.4	1.85	1.9

some measurement methods and measured variables are listed below:

- Profilometer (layer thickness)
- Four-probe measurements (dark conductivity, its activation energy)
- Two-probe measurements
- Transmission
- Reflection (surface roughness)
- Absorption (band gap)
- Spectroscopic ellipsometry (thickness, refractive index, layer sequences)
- Raman measurements (crystallinity, grain size, stress)
- X-ray investigations (see above)

6.1.3   Applications of Nanocrystalline Silicon

At present, a certain type of solar cell is attracting international attention in the area of the photovoltaic, i.e., the so-called heterojunction solar cell. In its simplest form, it is composed of a monocrystalline silicon substrate and an amorphous emitter deposited upon it. High absorption and stability of the amorphous layer as well as the quality of its interface to both the substrate and the antireflection coating still prevent mass production. Therefore, the amorphous layer is replaced by a nanocrystalline layer. The typical structure of such a solar cell is depicted in Fig. 6.2.

With such a solar cell, whose technological details are described elsewhere [141], Borchert *et al.* obtained an efficiency of 12.2 % in a first attempt [141].

6.1.4   Evaluation and Future Prospects

The only recognizable use of nanocrystalline layers are the above-named heterojunction (HIT) solar cells. In the meantime, however, Sanyo announced HITs on the basis of amorphous layers [142] specified with 21 % efficiency on a surface area of 100 cm^2. Thus, they will be able compete with manufacturers of other

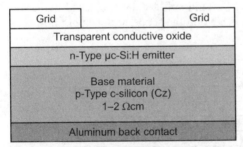

Fig. 6.2   Schematic representation of a heterojunction solar cell

cheap solar cells. It is evident that instead of amorphous layers, nanocrystalline layers are an area of research worth pursuing.

6.2   Zeolites and Nanoclusters in Zeolite Host Lattices

6.2.1   Description of Zeolites

One of the outstanding features of technological and industrial progresses in the last 30–40 years is the continuous miniaturization of countless technical devices and components and the resulting development of various procedures of micro and subsequently nanotechnological processes. The nanotechnologies in particular open possibilities which today are by far not yet foreseen in their varieties. The developments of these technologies have been and are from pronounced interdisciplinary character and can be found today practically in every area of high technology (beginning with microelectronics, optics, opto-electronics and sensor technology, numerous disciplines in chemistry, and in particular also pharmaceutics, medicine, biotechnology, etc.).

A strong impact for the development of micro and nanotechnologies arose from the necessity to sequentially reduce the geometry of electronic devices and integrated circuits in order to follow the demands for a higher complexity of circuits and devices. In particular the extreme reductions of the structural sizes in the context of the VLSI (very large scale integration) and ULSI (ultra large scale integration) technologies, with which modern computer chips are manufactured, were a strong driving force for this tendency. Today the vision of a global communication community drives the development of semiconductor technology. The handling of gigantic data quantities is required and therefore, there exists a continuously increasing demand for data storage of any type, faster and faster data transmission rates and more and more devices for monitoring and control.

The trend to ever decreasing structures is however not only limited to microelectronics. In the field of materials science the trend for miniaturization is also very strongly pronounced. In contrast to the continuous miniaturization in the microelectronics, which is driven by the necessity to reduce the geometry of individual devices and structures ever further in order to increase the component densities in complex microelectronic circuits, the reduction of the structural size is actually not the target in materials science. Rather within the field of the materials science the aim is the miniaturization within the nanometer range, because very small solids (nanoparticles or nanoclusters with a size of about 100 atoms) can show completely different material properties than the macroscopic solid state which is composed of the same atoms. This is to be attributed to the fact that nanocluster manifest a strongly increased surface-volume-relation in comparison to the macroscopic solid state. Dealing with smaller and smaller solids quantum effects become important. In addition it has to be emphasized that nanoclusters do not show molecular or atomic characteristics anymore. A substantial feature of the nature of nanoclusters is that their material properties partly depend on the number of atoms forming the cluster. Based on this peculiarities one can think about the development of materials with well defined and adjustable properties.

The developments of microelectronics indicate a fundamental approach to the implementation of extreme miniaturization and nanotechnologies, which is known as *top down*. The top down approach is based on a progressive reduction of the dimensions. The technologies used are based on lithography and sample transfer. Finally, dimensions down to about 10 nm are focused. Lately the so-called *soft lithography* methods turn out to be economical key technologies which could be used in different disciplines of the nanotechnology and are not limited to the microelectronic processing and procedures (like conventional methods such as soft x-ray lithography).

In contrast to the top down method one can also proceed from a *bottom up* method. This technological approach is based on the fact that individual functional elements of the nanotechnology are structured piece by piece from individual atoms or molecules. This approach is still to a large extent in the infant stage due to the obvious theoretical and technical difficulties. Nevertheless, today some interesting and promising applications already appear. For instance, one can think about pharmaceutical applications on the basis of molecule design or about nanostructures which are manufactured atom by atom with AFM devices (atomic force microscope). In the latter case one still cannot speak about functional elements or structures, but possible trends for future processing can be foreseen.

As a further principal method for the production of nanostructures or nanocrystalline solids, the use of porous materials which exhibit pores with open volumes in the order of 1 nm^3 can be manufactured. In this way nanocrystallites or nanoclusters which can also present volumes in the order of 1 nm^3 can be manufactured. A class of solids or minerals which meets the requirements of such form-giving framing materials is the family of the zeolites, since these manifest very porous crystal structures. In accordance with the various zeolite structures, numerous versions for the pore geometries can be found. Beside various cavities with a different geometry, quasi-one-dimensional channels with diameters in the nanometer range are also found. Since the cavities or channels are regularly arranged in the crystal structures of the zeolites, large areas of regularly arranged nanoclusters or bundles of quasi-one-dimensional structures can be created in zeolite host lattices.

6.2.2   Production and Characterization of Zeolites

Zeolites are a class of materials which manifest in their structure significant cavities, pores, and channels with diameters in the range of a nanometer [143]. These dimensions lie in the same order of magnitude as those of smaller molecules. Therefore zeolites can selectively adsorb molecules depending upon the size of the pore or channel diameter [144–146]. Industrially they have been used for decades as the so-called molecular filters or catalysts due to this characteristic. In addition, they occur as natural minerals, but on a large scale they are industrially manufactured since they particularly find a broad use in the petrochemistry. Practically all zeolites occurring in nature belong to the family of the aluminosilicates. In addition, many other zeolite compositions such as silicate, aluminophosphate, silicoaluminophosphate or titanosilicate can be synthesized.

The zeolite frame mostly has a negative (average) charge. Therefore the demand for a charge equilibrium of the total structure must be fulfilled by additional positively charged cations. These cations are often exchangeable. Exchangeable metal cations in dehydrogenated zeolites are, regarding their co-ordination, usually not saturated so that they can easily form complexes with different "guest molecules". This applies particularly also to organic molecules and works very effective if the guest molecules have a polar character or are not saturated [145].

The characteristic feature of the zeolites is their very porous structure which results from the fact that the crystal structures of the zeolites exhibit the form of interconnected polyhedrons (cubes, hexagonal prisms, so-called β-cages, etc.), whereby cavities and/or channels can occur in different sizes and shapes. The regularly arranged polyhedrons which define the structure of the individual zeolites are occupied at their corners with a silicon or an aluminum ion (this applies to the aluminosilicate) in each case. An oxygen ion is found in each case on the edges of the polyhedrons which connects neighboring Si and/or Al ions. In a zeolite four edges gather at each corner of the crystal structure with which the four valence of tetrahedrally coordinated Si and Al atoms is reflected. According to convention, the lattice atoms tetrahedrally coordinated in the zeolite structure (Si, Al) are referred to as T-atoms. Generally when one speaks about the silicates to which the zeolites are assigned, one also speaks about corner-related tetrahedron structures. The zeolites (or also more generally the aluminosilicates) manifest a crystal structure which with their network of interconnected tetrahedrons is very similar to the structure of silicon dioxide. The silicon and aluminum ions are exchangeable against each other in the structure, since the aluminum ions also manifest a co-ordination number of four. Still an open linkage must be saturated for the oxygen ions which connect aluminum and silicon ions so that the electrostatic valence rule is fulfilled. This saturation can take place via large monovalent or bivalent cations, i.e., via alkaline or alkaline-earth ions. For each aluminum ion built into the lattice an alkali or a "half" alkaline-earth ion is needed. In all zeolite crystal structures the ratio of the number of the oxygen ions to the total number of silicon and aluminum ions is 2:1. This relation follows inevitably from the structure of a complete tetrahedron network [147].

As example the zeolite types "Analcime", "Linde type A" or "Chabazite" with the chemical formulas $|Na_{16}(H_2O)_{16}|$ $[Si_{32}Al_{16}O_{96}]$, $|Na_{12}(H_2O)_{27}|_8$ $[Al_{12}Si_{12}O_{48}]_8$ or $|Na(H_2O)_3|$ $[AlSi_2O_6]$ or the aluminosilicate "Leucite" with the chemical formula $(K,Na)AlSi_2O_6$ used with these structures (in the last example potassium or sodium can be built into the structure) are considered. Like seen from these examples, an alkali ion per an aluminum ion is indeed found and the number of the inserted oxygen ions is twice as large as the combined number of Si and Al ions. It is observed that the cations (i.e., the alkali or alkaline-earth ions) and the water molecules in the zeolite lattices take defined sites and hence are of significant importance for the characteristics of the zeolites [147, 148].

Generally crystals which are structured like the zeolites or which have structures similar to the addressed tetrahedron networks possess some remarkable characteristics. So the alkali or alkaline-earth ions for instance, can partly be replaced by other ions in a solution (ion exchange). Because of these characteristics zeolites are used for example as water softeners in washing powders [147, 148]. The possibility for the ion exchange leads to the fact that the zeolites are used in particular in many catalytic processes. In this connection the ratio of the internal surface to the volume of these porous structures is very advantageous.

A good outline of the zeolites and their structures can be found in the Atlas of the Zeolite [149], which is accessible on the internet [150]. There, a good outline of the conventional notations and designations as well as a good overview of the structural setup of the individual zeolites can be found.

Production of Zeolites

The production of zeolites belongs to the standard processes in chemistry which are also extensively executed in the industry. Thus, it is referred to in extensive relevant literature, published in various journals, conference reports or monographs. A detailed description of the different manufacturing processes for zeolites goes by far beyond the scope of this short compilation. An outline and an overview of the production of zeolites can be obtained in [151–153].

It is only briefly mentioned here that the synthesis of large zeolite crystals is very difficult although they can quite occur in natural minerals as large single crystals. Artificial zeolite crystals are usually made of powdered or colloidal raw materials. First successes are obtained with the zeolite LTA directive ("Linde type A") and Na-X ("FAU", same structure as the mineral faujasite) [154, 155]. These crystals have diameters of approximately 65 µm (LTA) or approximately 140 µm (FAU). For many applications, large and cleanly grown single crystals are required. A procedure which is close to the assumed natural formation of large zeolite single crystals is presented in [155] (formation by static processes from compact materials). Different zeolite types are pulled in autoclaves at temperatures of 200 °C over long periods of time (up to 46 days), whereby crystal diameters of several millimeters are achieved [155].

Characterization of Zeolites

In principle, the description and characterization of zeolites take place with the usual standard characterization procedures in chemistry: x-ray diffraction (XRD), scanning electron microscopy (SEM), nuclear magnetic resonance (NMR), infrared (IR) absorption spectroscopy [151]. Furthermore, the capacities of the adsorption ability and the ion exchange are frequently used for the chemical characterization of zeolites [151]. Moreover, different physical or chemical characterization methods such as the atomic force microscopy (AFM), Raman spectroscopy, and high-resolution transmission electron microscopy (TEM) can of course be used.

The x-ray spectroscopy, particularly the powder diffractometry, is to be regarded as the standard method for the identification and characterization of newly synthesized zeolites. Therefore, it is most frequently used in laboratories which deal with zeolites [156]. From the position of the x-ray lines the dimensions of the unit cell can be derived. Lines which are not indexed manifest crystalline impurities or an incorrect indexing. If x-ray lines are systematically missing symmetry statements can be derived. From the intensities of the x-ray lines the structure of the sample can be determined (this analysis is very critical, therefore an accurate adjustment of the system with consideration of the specific conditions of the individual samples is absolutely necessary). Further information concerning possible portions of amorphous phases can be obtained from the background of the x-ray spectra. If the x-ray lines are further widened, information can also be received, for instance, about stresses, dislocations or the like in the crystal.

With regard to crystalline form and surface topology, zeolite crystals can be examined with the SEM whose magnitude covers the entire range from 20 nm to 100 μm [157]. Detailed evidence about the type of zeolite, aspect ratio of the crystal, distribution of the crystal sizes, twinning, surface roughness, etc., can be directly observed. Furthermore information about the homogeneity of a given set of zeolites or the amorphization can be gained. New substances might also be discovered. If a high-resolution transmission electron microscope is used for the investigations, then nanoclusters or quasi-one-dimensional structures also formed in the open pores or channels of the zeolite structures can be made visible and analyzed.

The NMR spectroscopy is a very important characterization method for the zeolite research [158]. The NRM in solid states is a technique which can be used as a complementary method to x-ray spectroscopy. The solid state NMR in contrast to the liquid NMR indicates some specific unique features and difficulties. In the following we will henceforth refer to the solid state NMR (even if we speak only of NMR). Both single-crystal samples and powdered or amorphous materials can be examined with the NMR. While the x-ray spectroscopy (preferably at single crystals) supplies statements about long-ranged orders and periodicities, the NMR permits investigations of the short-ranged order and structure. This makes the NMR today a valuable and well established method for revealing the structure of examined materials and in studying for instance, catalytic processes or the mobility characteristics of ions in the crystal. The potential which the NMR offers is already well-known for a long time. However, it is not trivial to detect solid state NMR spectra with the necessary resolution. In principle, the fine structure of the NMR spectra is lost when measurements are done on solid state samples since they often have strongly widened NMR lines so that substantial information for an accurate analysis of the spectra is lost. A reason for this can be the anisotropy chemical shift in the solid state (chemical shift: for an atomic nucleus different resonance frequencies can be expected, which can be assigned to different types of linkage with different chemical environments [159]). Besides, dipole and quadrupole interactions can become clearly apparent in the solid state since the molecules are not as mobile as in liquids. During the last years, however, some techniques have been developed which can suppress the disturbing interactions and phenomena [159] so that it is also possible today to obtain sufficiently well resolved NMR spectra from solid state samples.

All relevant atomic nuclei which are built into the frame structure of the zeo-lites (the so-called *framework*) can be detected with the help of the NMR (i.e., ^{29}Si, ^{27}Al, ^{17}O, ^{31}P). The natural abundance of ^{27}Al and ^{31}P lies within 100 %, therefore, the appropriate NMR spectra can be measured with good, (i.e., short) measurement times. However, ^{27}Al manifests a quadrupole moment which can lead to a widening of the NMR line by interacting with the electrical field gradi-ents. NMR analysis requires an enrichment of oxygen with the ^{17}O isotope since it naturally occurs only in very small quantities (0.037 %) [158].

The NMR lines of ^{29}Si and ^{31}P are normally narrow. These two elements (be-side ^{27}Al) play an important role as framework atoms in the zeolite structures. The ^{29}Si and ^{31}P NMR lines are very often used for the analysis of zeolites. The im-portance of the ^{29}Si NMR is based on the fact that the sensitivity of the chemical shift of ^{29}Si correlates with the degree of condensation of the Si-O tetrahedrons, i.e., the number and the type of tetrahedrally coordinated atoms which are bonded with a given SiO_4 complex. The signal of the chemical shift of ^{29}Si in $^{29}Si(n\,Al)$ with n = 0, 1, 2, 3, 4 (number of aluminum atoms which share any oxygen atoms with the concerned Si-O tetrahedron) covers a range from −80 to −115 ppm. The highest signal occurs for n = 0, i.e., if no aluminum atom shares oxygen atoms with the Si-O tetrahedrons. An important measure that can be obtained in the long run in this way is the Si:Al ratio of the zeolite frame. The existence of the so-called extra framework aluminum atoms can be proven by the ^{27}Al NMR, i.e., Al atoms which exist in the investigated structure in addition to those tetrahedrally built into the zeolite frame. Regarding catalytic applications in particular it is of great significance that the important dealumination process be pursued with the ^{29}Si and the ^{27}Al NMR [158].

In this connection it can be mentioned that techniques of solid state NMR can be developed for protons (^1H NMR), OH groups, adsorbed water, organic adsorb-ers, or for probe molecules, which again contain water molecules. The reason for this is to analyze the various state forms of hydrogen in the zeolites, for example, SiOH groups at which open linkages are not saturated by hydrogen (alkaline), AlOH groups of Al atoms (extra framework Al) which are not built into the frame of the zeolites, bridge-formed (alkaline) hydroxyl groups [SiO(H)Al], etc. [158].

It can be additionally mentioned that ^{129}Xe is a very suitable isotope for the analysis of the architecture of the pores and/or channels of the zeolites using NMR. The widely expanded electron shell of the heavy Xe inert gas atoms can be easily deformed by interactions with the pore or channel walls so that clear shifts are to be observed in the ^{129}Xe NMR lines from which conclusions about the pore or channel architecture can then be made [158].

A simple experimental method used to characterize zeolite structures is given by the measurements to the sorption capacity [160]. However, the data which are gained from the sorption capacity measurements permit only a qualitative estima-tion to the sample purity. These data do not allow any distinction of the various zeolite structures. It can only be measured whether the observed results are con-sistent with a zeolite structure that is already well-known [160].

For adsorbents with micro-pores, i.e., the zeolites, the equilibrium isotherm of the adsorption in a certain temperature range indicates a defined saturation limit which corresponds to a complete filling of the pores. At a constant temperature and with complete filling of the pores, the molecular volume of an adsorbed gaseous phase is very similar to the volume which corresponds to a liquid phase in the pores. Thus, from the measured saturation capacity of the adsorption a specific volume of the micro-pores can be measured. If the crystal density is well-known, then the portion of the pores in the total structure can be determined [160].

There are different methods used to determine the capacity of the adsorption ability [160]. With the so-called gravimetric method the sample which is to be examined is degassed on a micro-balance in vacuum (the sample is heated in the vacuum to higher temperatures and subsequently cooled down to the measuring temperature). Gradual quantities of the gas to be adsorbed are then let into the vacuum chamber and pressure and size modifications are recorded. Care must be taken before the measurement that the sample is really carefully degassed and that no residues of organic material remaining after sample synthesis are contained in it. Typical zeolites (e.g., ZSM-5) survive temperatures between 500–550 °C for some hours without structural damage and can therefore be oxidized at these temperatures in order to eliminate organic residues. Actual degassing occurs at 350–400 °C. These low temperature procedures can partly be compensated by long annealing times and a better vacuum. In principle, Al-rich zeolites have a small hydrothermal stability, i.e., their structure becomes easily unstable if they come in contact with water.

Probe gases which are to be adsorbed by the structures under examination can practically be all gases whose molecules (or atoms of noble gases) are not too large. Typical representatives are Ar, N_2, and O_2 to name a few. Also some paraffins (n-hexane) are flexible in such a manner that they can effectively fill out the pores of zeolites. Other molecules (e.g., i-butane) do not fill out the pores very well and therefore deliver too small values for the pore volumes. However, Ar, N_2, and CO_2 cannot penetrate the 6-oxygen rings, so that only volumes of pores whose entrance openings are formed by at least 8-oxygen rings can be recorded. The water molecule is also a very small molecule; besides, it forms a very strong dipole. Therefore, it is particularly strongly adsorbed by aluminum zeolite structures (on the other hand dealuminated zeolites are rather hydrophobic). In particular water molecules can penetrate into regions of the zeolite frame for which Ar, N_2, and O_2 are not accessible (e.g., in the so-called *sodalite cage*). From the comparison of the saturation capacities by the adsorption of different probe molecules qualitative structural information can then be indirectly derived [160].

Apart from the adsorption behavior of zeolites, ion exchange is also of prime importance [161]. This particular applies to the catalytic characteristics. If one proceeds from the classical zeolites which belong to the family of the aluminosilicates, the capacity of the ion exchange is given by the degree of the isomorphic substitution in the tetrahedron network, i.e., by the exchange of Si by Al ions [161]. Therefore, the theoretically possible ion exchange capacity is given by the elementary composition of the appropriate zeolite structure. The most sensitive analytic method for the analysis of the ion exchange is given by the use of radio isotopes, with which modifications in the composition of the frame structure can be easily proven. This occurs in particular with the help of the radio isotopes of the elements Na, K, Rb, Cs, Ca, Sr, and Ba [161].

To conclude this compilation of the most important methods used in the characterization of zeolites, the IR spectroscopy will be dealt with briefly [162]. Oscillations of the zeolite frame create typical bands (vibration modes) which can be measured with IR spectroscopy. These modes are situated in the middle and far infrared range of the electromagnetic spectrum. Originally the classification of the most important IR absorption modes fell into two groups, i.e., into internal and external vibration modes of the SiO or AlO tetrahedrons in the zeolite frame structure [162]. The following regulations are made in relation to the internal connections of the frame structure: asymmetrical stretching modes (1250–920 cm^{-1}), symmetrical stretching modes (720–650 cm^{-1}), TO bending modes (500–420 cm^{-1}). Related to the external connections are: the so-called *double ring vibration* (650–500 cm^{-1}), oscillations of the pore openings (420–300 cm^{-1}), asymmetrical stretching modes (1150–1050 cm^{-1}), symmetrical stretching modes (820–750 cm^{-1}). The spectral positions of the IR modes are often very sensitive with regard to structural changes. The initial classification into internal and external tetrahedron oscillations is not strictly kept and has to be modified [162]. In principle, the strict separation of the IR modes cannot be held since the individual oscillations are coupled together in the frame structure of the zeolites. Systematic modifications in the IR spectra are observed if for instance, the Al content in the tetrahedron network is varied. Thus, if necessary, the Si:Al concentration ratio in the frame structure can be analyzed using IR spectroscopy. Moreover, cation movements, for instance, can also be observed (e.g., during dehydrogenation) [163].

Raman spectroscopy is rarely used to analyze zeolites because it is often not simple to measure Raman spectra on zeolites with a sufficient intensity and an acceptable signal to noise ratio [164]. This is because of the loose frame structures of the zeolites. The Raman effect is generally a weakly pronounced phenomenon and hence the Raman spectra of zeolites are usually superimposed by a strong and broad background luminescence. In essence, two causes are identified for this background luminescence ([164] and references specified therein). Small quantities of strong luminous aromatic molecules can be available in the zeolite samples and cause the luminescence. These aromatic molecules are residues of organic raw materials which frequently remain in the zeolite samples as impurities after processing. Often this problem can be eliminated by a high temperature treatment in an oxygen atmosphere (but not always since the luminescence is sometimes even strengthened by the O$_2$ thermal treatment because organic molecules can possibly be transformed into a fluorescent phase). Moreover, Fe impurities in the zeolite samples can lead to a strong background luminescence. In principle, this problem can be avoided by performing highly pure synthesis procedures (however, this does not always hold for industrial mass productions). By Fourier transform (FT) Raman spectroscopy with excitation in the near IR regime the background luminescence is reduced as well. A detailed overview of the Raman modes observed in zeolites is given in [164].

6.2.3   Nanoclusters in Zeolite Host Lattices

Nanocrystalline materials which are also called nanoclusters or nanoparticles can clearly manifest deviations in relation to their "normal" macroscopic physical states. This can apply, for instance, to their optical, electronic, or thermodynamic characteristics. For example, in nanocrystalline Sn clusters a shift in the melting point as a function of the particle size can occur. In strongly porous crystal structures, as they are manifested by zeolites with open pore volumes of 30–50 %, nanocluster can be formed from various materials. Here, the zeolite frame serves as a designed frame structure. Since the open pores can be present in different well defined crystallographic geometry in the numerous zeolite structures, theoretically one can directly manufacture evenly structured nanoclusters from different materials with various particle sizes. This prospect opens a further field of possible applications. For example, molecular filters for various chemical process cycles through which storage of problematic nuclear wastes can be achieved in the framework of nuclear waste management up to the establishment of future nanoelectronic devices or computers. However, the latter examples are still far fetched and presently, a matured product is still to be settled in the area of the scientific visions. Nevertheless, numerous fundamental and promising scientific material statements have been developed.

Production of Nanoclusters in Zeolite Host Lattices

Different techniques are developed in order to synthesize and stabilize metallic and semiconducting particles or nanocluster with geometrical dimensions on the nanometer scale. In order to control the size and distribution of the nanocluster, zeolite with their numerous versions of pore geometry and distributions offer very suitable host lattices for the production of various large arrangements of nanoclusters [165–177].

It is noteworthy that there is the possibility to produce definite individual nanoparticles in the *confinement* of a zeolite pore (cage) and to regularly arrange them simultaneously in greater numbers due to the given crystal structure of the host lattice. Ideally, a field of identical nanoparticles which are arranged in a *superlattice* is then obtained. Thus, a material which manifest the characteristic of a nanocluster (e.g., the ability to emit light which in relation to the macroscopic solid state of the same material is blue-shifted) is achieved. Due to the immense multiplicity of the clusters arranged in the superlattice this microscopic characteristic can then be used macroscopically.

The production of nanoclusters in the zeolite host lattices can be implemented for various metals such as Pt, Pd, Ag, Ni, semiconducting sulfides, and selenide of Zn, Cd, and Pb or oxides such as ZnO, CdO, SnO_2 ([168] and references quoted therein). The host lattice works like a solid state electrolyte. In solutions or melts mobile cations which compensate the charge (e.g., Na^+) by mono and multivalent cations are exchanged and are then reduced by suitable substances such as hydrogen. These processes require the mobility and agglomeration of metal cations or atoms which spatially occur separately before the reduction since they sit on defined cation sites. Unfortunately, the formation of nanoclusters leads in many cases to a local disturbance or degradation of the host lattice (e.g., by local hydrolysis of the zeolites). As a consequence, the previously well defined pore sizes and concomitantly the sizes of the formed nanoparticles are changed. Under this circumstance, the nanoclusters are no longer present as homogeneous particles. Thus, the confinement for the size adjustment is softened or at worst even removed [168].

The production of CdS nanoclusters in a zeolite-Y host lattice is described in [167, 178, 179]. Zeolite-Y appears in nature as the mineral faujasite and consists of a porous network of Si and Al tetrahedrons which are connected by oxygen atoms [165]. Thus, zeolite-Y has a frame structure which is typical for aluminosilicates. Two sorts of cavities are formed by its frame structure: (i) the sodalite cage with a diameter of 0.5 nm which is accessible to molecules by a circular window of 0.25 nm in diameter, and (ii) the so-called *supercage* with 1.3 nm diameter and a window opening of 0.75 nm in diameter. These two cavities, with well defined sizes and arrangements form a suitable environment in which smallest crystalline clusters are formed. The participating ions of the reagents can be supplied through the window openings. CdS nanoclusters can then be synthesized by ion exchange in the zeolite-Y matrix [167, 179].

The production of various other guest clusters in a confinement of zeolite frames is also examined [167, 180, 181]. AgI is manufactured in the zeolite "Mordenite", and PbI_2 in X, Y, A, and L-type (Linde type) zeolite host lattices [167]. All these nanoclusters in host lattices clearly show changed optical characteristics in comparison to the "normal" behavior of a macroscopic crystal. CdS clusters could be implemented into different cages and channels of various zeolite host lattices [167, 182]. The size of the respective cluster is limited by those cages or channels. The CdS clusters are formed in the largest cages or in the main channels of the zeolite structures. Absorption spectra of the CdS clusters in the zeolite frame indicated two versions, which reflects the two different confinement types, i.e., cages and channels.

SnO_2 clusters are formed in a zeolite-Y matrix [183, 184]. This binding takes place by ion exchange in a $SnCl_2$ solution. The portion of Sn can vary between 1 and 11 weight per cent and the size and topology of the clusters depend on the Sn loading [167, 183, 184]. The cluster sizes cover a wide range between 2 and 20 nm diameter. The larger particles probably present secondary aggregates which are bonded together with smaller clusters [167, 185]. Here, the above mentioned softening of the frame structure is shown. This softening can lead to the fact that the cluster looses its well defined sizes.

Regarding the production of one or quasi-one-dimensional electrical conducting structures (1D nanowires) metal-loaded zeolites with suitable channel structures are suggested as promising candidates [168, 186, 187]. Thus, the dehydrogenated K^+ form of L-type zeolite, for instance, is loaded with different quantities of potassium [188, 190]. With rising potassium loading the conductivity of the material increases. The conductivity increases with rising temperature and is thus thermally and not metallically activated. It is questionable or even doubtful whether this method of producing quasi-one-dimensional conducting structures is a suitable way in the direction of the production of electronic devices on the nanometer scale [168]. In this connection, it is a problem that the individual channels are geometrically too closely packed together because separating neighboring conductive channels from each other and hence really ensuring a quasi-one-dimensional current conduction is difficult. Besides, the zeolite material loaded with potassium is very reactive and thus makes the handling of the substance and its application for future electronic functions problematic or impossible. Nevertheless, the study of such composite materials is of fundamental scientific interest and should be given further attention.

Characterization of Nanoclusters in Zeolite Host Lattices

The characterization of nanoclusters in zeolite host lattices can take place with different methods. A very direct method is of course the transmission electron microscopy (TEM) or generally the high-resolution electron microscopy (HREM). In this connection, a very detailed outline article has been published in 1996 by Pan [191]. In the article, the meaning of HREM methods for the zeolite research is discussed and it also deals in particular with the analysis of nanoclusters in the zeolite host structure. The article [191] gives a global outline of the special HREM techniques for the characterization of zeolite structures. However, the analysis of zeolites or nanoclusters in the pores of the zeolite structure is not completely unproblematic, since the open zeolite frame structures are rather unstable with regard to high-energy electron radiation. Consequently, the possibilities of HREM with respect to structural analyses in zeolites and hence the investigations of nanoclusters have been somehow limited up to recently. Downwards, the maximum resolvable structures are limited to approximately 0.3 nm. In the last years the progress obtained with the development of the so-called *slow scan CCD systems* (charge-coupled device) has created room for improvements since beam performances can be reduced with the same resolution (low dose image).

HREM investigations have been published for more than 20 years regarding the formation of nanoclusters in zeolites. The emphasis has been firstly laid mainly on small metal particles since these are of great importance for catalytic processes in the petrochemistry (e.g., [192, 193]). Later semiconducting nanocluster were then of interest (e.g., [194]), which became more important in the context of the investigations of quantum dots. HREM investigations have been executed essentially in order to study, for instance, the distribution of particle sizes (e.g., regarding the correlation between structure sizes and function/efficiency of metal catalysts).

Furthermore, the local positions of the metal clusters in the zeolite frame with regard to their formation and their growth are of interest. The third important information which can be clarified with HREM methods is the relationship between the zeolite host matrix and the particle structure.

Analyses of the optical properties have been proven as further very important and frequently used methods to obtain information about the characteristics of nanoclusters in zeolite host lattices. This applies largely to the study of semiconducting nanoclusters such as CdS (e.g., [195]). In [195] for example, CdS nanoclusters which are synthesized in the pores of different zeolite hosts are optically analyzed (luminescence, i.e., excitation and emission spectra, optical absorption, etc.). The spectral shifts (blue shift) always observed in nanoparticles are explained in the context of the QSE (quantum size effect) model. Similar investigations concerning CdS, Ag, Cu, AgI clusters and nanoclusters in zeolite-Y samples [179, 196–198] are also published by other authors.

Raman spectroscopy offers a further possibility of examining nanoclusters [164, 199, 200]. Adsorbed molecules or various metal complexes in zeolite frames are examined (see the outline article [164]). Raman studies of Se, RbSe and CdSe clusters in zeolite-Y have shown that these nanoclusters manifest similar characteristics as the disturbed bulk phases [199]. The authors of [200] investigated chalkogenides introduced into the pores of zeolites by Raman spectroscopy and came to a similar conclusion. Here the Raman spectra of amorphous, glass-like a-$As_{22}S_{78}$, bulk samples and AsS nanoclusters in a zeolite matrix (zeolite A) manifest great similarities.

A further method which can be used in the analysis of nanoclusters in zeolite host lattices is the thermal-gravimetric method (or microbalance thermal analysis, TA), which permits the investigations of adsorbed molecules in zeolite structures as a function of the temperature [201].

Detailed x-ray powder diffractometry and EXAFS analyses (extended x-ray absorption fine structure studies) can also be employed in the analysis of nanoclusters [179]. However, these analytical methods are very complex.

6.2.4 Applications of Zeolites and Nanoclusters in Zeolite Host Lattices

Like already mentioned, zeolites are used for several chemical applications. This applies in particular to industrial applications in the proximity of catalytic functions [143, 144, 165]. One of the most important applications is the use of zeolites as diaphragms which is based on its characteristic as molecular filters. A good overview to this topic can be found in the outline article of Caro *et al.* [202]. Ideal zeolite diaphragms combine the advantages of inorganic diaphragms, i.e., temperature stability (in principle up to 500 °C) and dissolution resistance with an almost perfect geometrical selection behavior. The latter characteristic is of course linked with the various pore and channel geometries which can be found in the various zeolite types. The importance of zeolite diaphragms for the industry becomes clear from statements from different studies (see [202] and references quoted therein) that a current market volume of approximately 1 billion US$ with simultaneous growth rates of 10 % is predicted (for year 2000 [202]). In various research and development activities which have been carried out lately regarding inorganic diaphragms (and still continue), zeolites are of significant interest beside micro-porous diaphragms which are based on sol gel processes and Pd-based diaphragms.

A further current area of application for zeolites are the so-called *zeolite modified electrodes* (ZMEs) for the electro-analytic chemistry [203]. The attractiveness of the ZME is based on its capability to combine the ion exchange capacity of the zeolites with their selection abilities on the molecular scale (molecular filters). Here, numerous promising analytic or sensory applications appear but will not be further discussed here. (The reader is referred to the outline article [203].)

A further field which should also be mentioned here only briefly, is the use of zeolites as media for the storage of hydrogen (see e.g., [204]). Applications are with regard to a safe fuel storage for hydrogen-operated vehicles or in the case of hydrogen transport.

A completely different promising field of application for zeolites is found in the area of luminescence materials or phosphors for various luminous technical applications (solid state luminescence) [205]. Here in particular, there are immense possibilities if the modification of the luminescence characteristics of zeolites by the installation of nanoclusters in the zeolite frame is considered.

The trend towards ever growing miniaturization in electronics in the direction of nanotechnology will sometime necessitate the development of radically new technological procedures. If the focus is on quasi-one-dimensional operating electronic devices or current conductors, the chances for success in the context of the current existing technologies are few [187, 206]. Perhaps a long-term perspective offers a completely new concept which is referred to as *crystal engineering* [207] for the production of such devices [208]. The vision is that inorganic materials be completely designed on the nanometer scale, whereby in the long run the aim of producing a material with a band structure adapted for a certain application will be achieved. For instance, with reference to semiconductors one can speak of a *band gap engineering*. In this connection, zeolites which are loaded in their channels with metal clusters are constituted as possible candidates for the production of closely packed, quasi-one-dimensional electrical conductors [208].

Initial investigations are already executed in this direction. However, they still move intensively on the level of fundamental material research and show some perspectives at best [208]. Dehydrogenated zeolites (e.g., of the L-type) with which cations are coordinated to an anionic frame only on one side form the insides of regularly arranged channels. A continuous doping of the normally isolating zeolites with excess electrons is possible by a reaction of the zeolites with metallic alkali atoms (from a gaseous phase). The alkaline metal ions are ionized by the strong electrical fields within the zeolite structure so that electrons which can interact with the cations of the zeolite structure are set free [208–214]. An intensified electron-electron interaction and the possibility of an insulator-metal transition for the zeolites starting from a critical loading of the channel/pores with metals can be expected [208, 214–216]. Some promising experiments are presented in [208], where clues about an anisotropic electrical conductivity are found after potassium doping of the channel structures of L-type-zeolite (by eddy current loss and electron spin resonance measurements, ESR).

6.2.5   Evaluation and Future Prospects

Like already mentioned several times, zeolites have an important position in the chemical industry due to their various applications particularly regarding catalytic processes (e.g., in the petrochemistry). This will not change in the near future if the growth prognoses for zeolite diaphragms is considered (see Sect. 6.2.4 and [202]). The use of zeolites for sensory assignments and of course particularly for chemical sensors is also promising and will be probably developed in the future. By increasing sensitivity which concerns environmental aspects, growth rates are clearly to be expected.

There are at present no concrete applications especially in the area of electronics with regard to nanoclusters which are built into zeolite frame structures. However, on average there are some applications in the area of luminescence materials or phosphors. Here significant growth rates might be expected in the future (although with a certain risk), since the requirement of such materials will rise par-

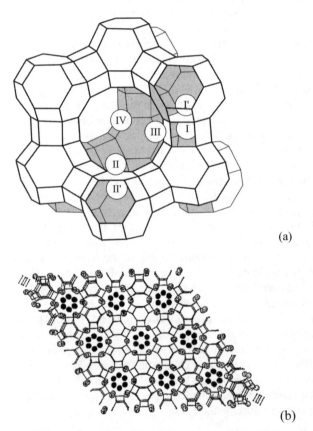

(a)

(b)

Fig. 6.3   (a) Structure of the faujasite (synthetically also zeolite-Y) [205]. In the center the so-called *supercage* can be seen (see also [143, 144, 149, 150]). Like easily seen, the lattice of this zeolite structure is formed from two basic elements. The position of the oxygen ions in the frame (•) and the position of the cations (I, I', II, II', III, III') are sketched. (b) Schematic example of the clustering of potassium ions (•) in the channel structure of zeolite-L [208]

ticularly under the point of view of energy conservation measures. Comparatively, electronic applications regarding nanotechnology and electronics with quasi-one-dimensional current transport lie rather in the distant future (Sect. 6.2.4).

7 Nanostructuring

7.1 Nanopolishing of Diamond

7.1.1 Procedures of Nanopolishing

Grinding, thinning, beveling, and polishing are the first steps to shaping and structuring a material. At first sight, these methods appear to be simple. However, some materials would offer interesting applications if they could be processed mechanically. Such an example is diamond, which is treated in this section. The special interest in diamond is fine polishing for optical applications and the production of blades for surgical tools by beveling.

For natural diamonds or artificial ones manufactured by high pressure and high temperature, the problem of polishing has not been resolved economically, but at least technically. The stones are sharpened and polished between two rotating cast iron plates using diamond powder as an abrasive. It is possible to sharpen the stones up to a roughness of a few nm. However, there is a risk of breaking the beveled edge with high pressures. A further disadvantage is the large anisotropy of polishing in the various crystallographic orientations. It is almost impossible to polish the crystal in the (111) orientation.

For economic reasons—price, assurance of a constant supply and quality—it is desirable to replace the monocrystalline diamond with polycrystalline films. According to their manufacturing process from the gaseous phase, they are called CVD films *(chemical vapor deposition)*. Since the crystallites composing the film are arranged randomly, there are always some that are aligned in the diamond's hard direction. Instead of being polished, these ones are rather torn off the surface. Thus, the roughness of the surface increases, and gaps appear at the edges.

Several alternatives were investigated to overcome these problems, e.g., etching in molten rare earth metals or transition metals, sputtering with low-energy ions, solid state oxidation, among other things. The pros and cons of these methods are discussed in [217] and the literature quoted therein. Substantial restrictions turned out in each case.

Some years ago, a method was developed that is deemed most promising today: thermochemical polishing [217–219]. Its setup is presented in Fig. 7.1. A diamond sample is placed on a rotating plate made of a transition metal, e.g., iron. A second plate (weight)—made of the same transition material—is placed on this sample. The chamber in which the polishing takes place is heated to a temperature between 700 and 1200 °C (high temperatures lead to rapid but rough polishing, while the extremely fine polishing takes place at moderate temperatures).

Polishing is supported by different measures like the mechanical vibration at the rotation axis and the inlet of an argon-hydrogen gas mixture.

The mechanism of thermochemical polishing is based on the conversion of the diamond surface into graphite. This is a well-known process which occurs, for instance, when diamond is annealed after ion implantation. It is substantially accelerated by the selective contacts between the diamond and the transition metal. In the second step, the graphite formed diffuses into the transition metal. Therefore, a transition metal of low carbon content is used and the polishing plates are changed after some time to avoid the saturation of carbon. Hydrogen works as a catalyst. This concept can be described by a model and is treated mathematically in [220].

7.1.2   Characterization of Nanopolishing

For the optimization of the procedure, the etching rate is measured as a function of different parameters like temperature, pressure, angular velocity, vibration frequency, vibration amplitude, etc. [219]. In Figs. 7.2–7.7, some results are shown in order to give an idea of the etching rates that can be achieved.

The following values apply to the above-named figures: sample diameter 10 mm, temperature 950 °C, mass 11.704 g, angular frequency 112 cycles per second, vibration frequency 450 cycles per second, and vibration amplitude 3.46 mm (with the exception of the variable parameter).

The illustration of the removal rate can be improved by an Arrhenius plot (Fig. 7.3). Two activation energies of 1.42 and 0.52 eV are measured. However, a theoretical model explaining these data is still missing.

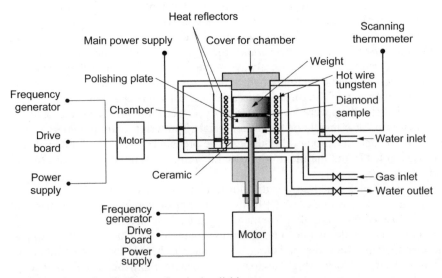

Fig. 7.1   Setup for the thermochemical polishing

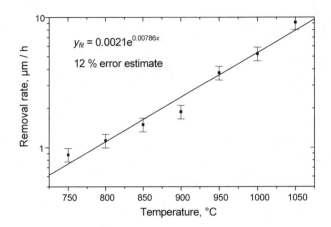

Fig. 7.2 Removal rate as a function of temperature

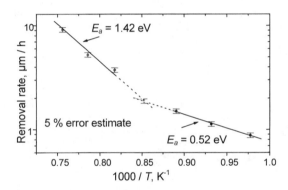

Fig. 7.3 Arrhenius plot of the removal rate

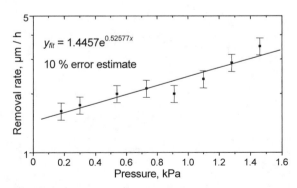

Fig. 7.4 Removal rate as a function of pressure

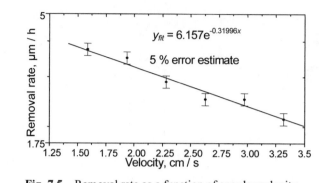

Fig. 7.5   Removal rate as a function of angular velocity

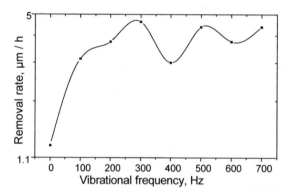

Fig. 7.6   Removal rate as a function of frequency of vibrations

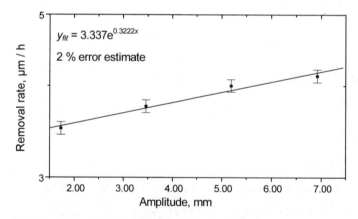

Fig. 7.7   Removal rate as a function of amplitude of vibrations

For the determination of the non-diamond carbon phases occurring during polishing, Raman measurements are performed [218]. The Raman spectra are seen in Fig. 7.8a–f. The spectrum of the as-grown film (Fig 7.8a) shows a small line at $1206\ cm^{-1}$ beside the diamond signal at $1331\ cm^{-1}$ and a broad peak from $1350\ cm^{-1}$ to approximately $1555\ cm^{-1}$. The line at $1206\ cm^{-1}$ is assigned to a mixture of sp^2 and sp^3 carbon. The broad peak is assigned to a common band of disordered (nanocrystalline) graphite with a peak around $1353\ cm^{-1}$ and amorphous diamond-like carbon with a peak around $1455\ cm^{-1}$. Figure 7.8b shows the spectrum after thermochemical polishing for sixteen hours. The broad peak in Fig. 7.8a now splits up into two peaks. The nanocrystalline peak at $1353\ cm^{-1}$ is a result of disordered sp^2 bonding, while the amorphous peak at $1455\ cm^{-1}$ originates from disordered sp^3 bonding. Further polishing (Fig. 7.8c) leads to the emergence of an additional peak at approximately $1580\ cm^{-1}$. This is the microcrystalline peak resulting from well-ordered sp^2 bonds. The bands of the disordered graphite and the amorphous diamond-like carbon are now closer together than beforehand. Further polishing results in the Raman spectrum of Fig. 7.8d. The diamond line and the band of the amorphous diamond-like carbon completely disappear. The latter is presumably converted into two graphite phases with a sufficiently thick layer that extinguishes the diamond line.

After further polishing for 16 hours at moderate temperature (800 °C, Fig. 7.8e) the nanocrystalline and microcrystalline phases are gradually washed away from the surface of the diamond down into the polishing plate by diffusion so that in effect only a flimsy trace of the nanocrystalline graphite band and a small microcrystalline band are visible. The diamond line shows up again, which signifies the reduction of the graphite layer to a value below 1 μm (the penetration depth of the Ar laser). Figure 7.8f shows the Raman spectrum after finally fine polishing at 750 °C and with moderate pressure. The intensity of the diamond line increases substantially; there are no more graphite portions. The position of the diamond Raman line at $1331\ cm^{-1}$ on the polished films is proof that thermochemical polishing does not impair the lattice structure of the surfaces.

A first result of nanopolishing is already shown in Fig. 4.42. At present, the minimum attainable surface roughness amounts to 1.2 nm. Furthermore, a typical beveled edge is shown in Fig. 7.9 (successive beveling of both the front and rear sides of a CVD diamond film) [221]. The radius of curvature of the beveled edge is approximately 50 nm.

7.1.3   Applications, Evaluation, and Future Prospects

Diamond offers several unsurpassed properties: highest mechanical hardness, highest heat conductivity, high stability against chemical attacks, high radiation inertness, no incorporation of impurities, and optical transparency. Diamond finds application when the combination of plane surfaces and any of these characteristics is required. Most probably this may be the case in the optical area. However, there is no demand in this area at the moment.

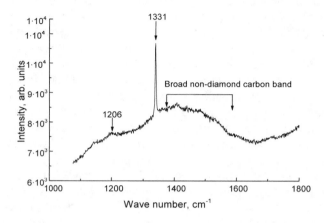

(a)

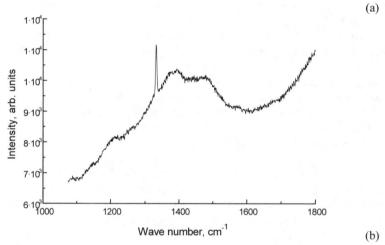

(b)

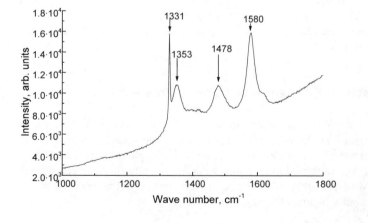

(c)

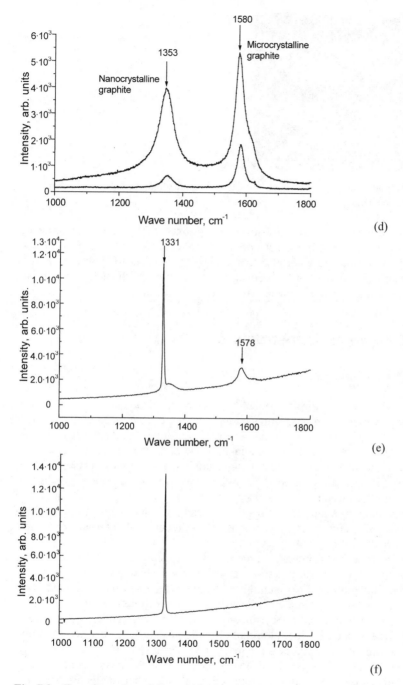

Fig. 7.8   Transformation of diamond into non-diamond carbon phases during thermochemical polishing

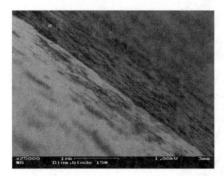

Fig. 7.9 Formation of a blade by successive beveling a diamond film on the front and rear sides of the same edge

Application regarding surgical instruments is conceivable in the case of beveling and the production of blades. An obvious application could be in the eye surgery. At the moment, there are attempts to shape nanotubes from diamond conically in order to be able to further convert it into pipettes. With such a tube, tissue samples or liquids could be extracted from areas of a few micrometers in diameter.

7.2   Etching of Nanostructures

7.2.1   State-of-the-Art

Apart from the common direct production of powdery nanomaterials for the surface coating by deposition procedures or agglomeration from molecules, the direct production of regular structures in nanometer dimensions of full surface deposited homogeneous layers by etching techniques is of highest interest. In microelectronics, micromechanics, sensor technology, and integrated optics, further fields of application for accurately defined nanostructures are, for instance, conductive strips in integrated circuits, gate electrodes of transistors, and optical gratings. The necessary structures cannot be attained by wet etching due to the isotropic etching characteristic of many reaction solutions. Consequently, alternative procedures must be carried out.

For applications requiring high precision in the geometrical dimensions, dry etching in the parallel plate reactor is a well-known technique from microelectronic circuit integration. This etching technique, which has been adopted since about 1980, enables a reproducible structuring of very different materials of semiconductor technology. In addition, silicon in crystalline and polycrystalline form, silicon dioxide, silicon nitride, aluminum, titanium, tungsten, polymers, polyimide, and photoresists among others are materials under investigation.

The basic structure of a parallel plate reactor for dry etching is presented in Fig. 7.10. One of the two electrodes is grounded, while the other one is adjusted to a high frequency (13.56 MHz). The procedures are grouped depending upon the coupling of the high frequency to the upper or lower electrode. If the layer to be

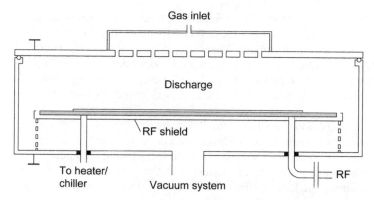

Fig. 7.10 Setup of a parallel plate reactor for dry etching in PE or RIE procedure

etched is on the grounded electrode, then we are dealing with plasma etching (PE). If it is on the HF coupled electrode, we are dealing with reactive ion etching (RIE).

In both ways, plasma excitation between the electrodes is used for material removal. Fluorine or chlorine-containing gas compounds serve as reaction gases, for instance, SF_6 or $SiCl_4$. Due to the radio frequency (RF) excitation, electrically neutral radicals, positively charged ions, and free electrons are generated between the electrodes. Due to their small mass, the electrons can follow the high frequency, but the slow-acting ions cannot. This leads to a negative charging of the RF fed electrode during the positive half-wave of the RF signal. Since the electrons cannot leave the electrode during the negative half wave, the electrode remains negatively charged in average.

The developed voltage is called bias voltage, and the resulting potential between the electrodes is presented in Fig. 7.11.

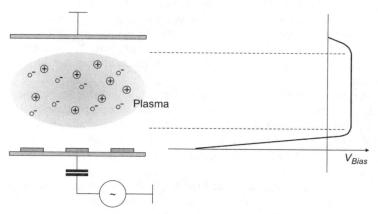

Fig. 7.11 Potential between the electrodes of the parallel plate reactor for reactive ion etching [223]

The bias voltage produces an electrical field, which now accelerates the positively charged ions to the RF electrode. Due to their kinetic energy, they knock out material from the surface upon impact. Therefore, besides the purely chemical material removal by the radicals which occurs at both electrodes, additional physical etching occurs at the RF electrode. Thus, plasma etching is a purely chemical, very selective etching, while reactive ion etching is a mixed physical/chemical etching procedure.

For many applications, both the PE and the RIE meet the demand of the etching rate and the selectivity between the materials. However, a sufficient anisotropy of the etching procedure is critical in the case of the plasma etching technique. Here the RIE etching technique clearly has advantages. Even under extreme conditions, e.g., 200 nm polysilicon on 1.5 nm silicon dioxide, anisotropic polysilicon etchings are performed with this procedure without destruction of the gate oxide [222].

Due to the low process pressure, the mean free path of the particles in the etching reactor is in the centimeter range so that the charged ions or molecules are accelerated strongly towards the charged electrode. On their way, they rarely collide with other molecules and thus, they do not deviate from their direction of motion. Therefore, these particles hit perpendicular onto the surface of the electrode, and an anisotropic etching process develops.

While the ions cause a directed physical material removal, the excited radicals lead to a chemical and extensively non-oriented etching. The extent of both etching portions determine both the degree of anisotropy and the selectivity of the etching process and can be influenced by the fed RF power and the pressure of the reactor.

The etching rate due to mechanical-physical etching grows with increasing high frequency power. Moreover, a decreasing pressure in the recipient leads to less impacts between the available gas particles, and thus, also to a rise in the average energy of the ions via an increase in the mean free path. Therefore, the ions hit almost perpendicular onto the substrate surface, and the material removal occurs anisotropic.

The etching rate due to chemical etching is essentially determined by the reaction gas used. Depending upon the material to be etched, fluorine, chlorine or rarely bromine and iodine compounds are used. Radicals are excited due to gas discharge and transfer the etched material into the gaseous state by compound formation. The reaction products are removed from the reaction chamber through the vacuum system.

At present, the RIE method is state-of-the-art in semiconductor technology. Here, SiO_2 and Si_3N_4 are removed in general by fluorine-containing gases (CF_4, C_2F_6, CHF_3), while etched aluminum, polysilicon, and crystalline silicon are removed by chlorine ($SiCl_4$, CCl_4). Particularly under high aspect conditions, heavier ions, such as bromine or iodine in the form of hydrogen compounds, are increasingly used during silicon etching.

The dry etching technique can also be used for the production of nanostructures from full surface deposited layers without major modifications. Here, photoresist films usually serve as masks. Materials, which are highly resistant to etching in relation to the layer to be etched and to the process gas employed, are also fre-

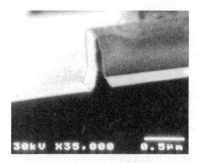

Fig. 7.12 RIE-etched crystalline silicon structure of 800 nm height and 80 nm width at the tip, masked with 100 nm silicon nitride

quently used. An example of a crystalline silicon structure etched by the RIE method with a 100 nm thick Si_3N_4 mask layer is shown in Fig. 7.12.

7.2.2 Progressive Etching Techniques

Further developments in reactive ion etching are inductively coupled plasma etching (ICP) and electron cyclotron resonance plasma etching (ECR). With reference to the energy of the excited radical ions, the independently controllable dissociation rate of the reaction gas via two separated high frequency generators is common to both procedures.

By this separation, high densities of reactive radicals can be produced despite a small operating pressure in the reactor because a high dissociation degree of the gas is achieved by means of a large excitation RF power of the plasma source. There is no influence on the particle energy. This is only determined by the bias voltage placed at the substrate electrode via a second RF generator.

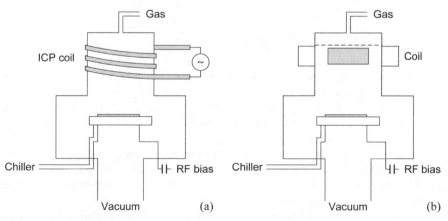

Fig. 7.13 Schematic cross section of the ICP (a) and ECR etching device (b), according to [224]

Thus, very high etching rates of up to about 10 $\mu m/min$ can be achieved due to the attainable high radical densities. Simultaneously, extremely high selectivity is given as a result of the small particle energy. Additionally, almost completely anisotropic material removal takes place due to the large mean free path of the radicals at the small process pressure.

The ICP etching technique finds increasing applications for micromechanical and deep silicon trench etching with high aspect ratios. The acceptance of this equipment also increases in the area of required high selectivities such as the structuring of polysilicon on thin gate oxide.

In the case of ECR etching technique, inhomogeneities occur in the plasma distribution due to resonance shifts in the source. This technique does not find much application in industry.

7.2.3   Evaluation and Future Prospects

Although the progressive procedures enable higher etching rates with simultaneous improvement of the selectivity, the results attainable with the reactive ion etching technique are basically still sufficient for many future applications. In the meantime, the microelectronics industry uses ICP etching devices for gate structuring in the production of new products with minimum dimensions within the deep submicrometer range, in order to get a larger process window with regard to the selectivity between the materials.

The inductively coupled plasma device is generally performed within the range of anisotropic depth etching because appropriate etching depths with conventional RIE systems are not attainable (cf. Fig. 7.14).

However, the throughput of this device is limited. Deep etching with aspect ratios above 20:1 requires a substantial amount of time. Additionally the maintenance expenditure of this device is quite high since sulfur deposits in the evacuated system lead to increasing wear.

7.3   Lithography Procedures

The term *lithography* generally means the transfer of structures of an electronic or an image pattern into a thin radiation-sensitive layer, the photoresist, by means of electromagnetic waves or particle beams. The execution of the lithography method involves a series process consisting of deposing the photoresist, exposure and development of the radiation-sensitive layer.

The photoresist deposition on the substrate takes place via spin-coating in which the resist is given on a rotating plate (approximately 3000 rpm). A homogeneous coating of the surface is achieved by means of the centrifugal energy in combination with the viscosity of the resist.

Alternatively spray coating which leads in particular to a higher uniformity in the boundary region of asymmetrical bodies is used for larger substrates.

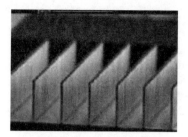

Fig. 7.14 ICP-depth etching with high aspect relation in crystalline silicon

As procedures for the exposure, optical, x-ray, electron and ion beam lithography of different versions are at disposal. All these mentioned techniques enable a reproducible, highly resolved structural production on the substrate coated with photoresist whereby the optical lithography manifests the smallest resolution because of the largest radiation wavelength.

In the lithography technique, developing the resist means removing the exposed or unexposed areas in a base solution. Development takes place in NaOH or TMAH solution by dipping. Alternatively, the spray development offers the highest reproducibility.

The subject of further sections is the transfer of the structures by irradiation of the resist, generally known as the exposure procedures, as well as the respective procedures belonging to the mask technique.

7.3.1 State-of-the-Art

In the research areas of universities, the economical suitable optical contact lithography with UV light is used which enables a resolution in the upper submicrometer range, but with reduced yield. However, semiconductor manufacturing plants and research institutes use the expensive projection exposure as wafer scan, step and repeat or step scan procedure which also enables a small defect density and thus a high yield, beside the improved resolution. Electron-beam writers are used for mask making and sometimes for direct substrate exposure, too.

7.3.2 Optical Lithography

Today the optical lithography with light in the wavelength range of 465 nm down to 193 nm is used for the structural transfer from the mask onto the photoresist in all micro techniques for production. Also in the research lab the optical lithography in contact mode is very common.

Contact Exposure

The contact lithography uses masks from glass on whose surface the desired structures are available on a 1:1 scale in the form of a thin chromium film as absorber. Boron silicate or quartz glass is used depending on the selected wavelength.

During the contact exposure the photomask is in direct contact with the photoresist film at the surface of the substrate so that during irradiation of the mask the structures are transferred on a 1:1 scale. For the contact improvement of the resolution the substrate is pressed against the mask before the exposure. Additionally, vacuum is applied between mask and substrate.

The resolution is limited only by the diffraction effects at the structure edges so that minimum structural widths of about 0.8 μm for 436 nm wavelength down to about 0.4 μm for 248 nm wavelength are possible on plane surfaces as a function of the photoresist thickness and the used wavelength [225]. By decreasing the wavelength to 220 nm line widths of about 100 nm are obtained [226]. Presumably, the procedure can also be extended to structural widths below 100 nm by further reduction of the wavelength.

An obstacle for the application of the contact exposure in nanotechnology is the production of extremely fine structures on the masks. On the one hand, the writing of the 1:1 masking is very time consuming and thus expensive for these structure widths due to the substrate size mask surface. On the other hand, extremely thin photoresist films are required for the suppression of the diffraction influence at the structure edges.

Since all chips of a substrate are exposed simultaneously with a 1:1 mask, a high throughput is possible with the contact exposure. The exposure devices are less expensive and maintenance is not intensive. However the masks are relatively expensive.

The disadvantage of this procedure is the unavoidable position-dependent adjustment error of already manufactured structures on the substrate, resulting from temperature gradients and mechanical stresses, as well as the strong load of the expensive mask by direct contact between mask and substrate surface. The contact leads to a fast contamination of the mask, possibly existing particles between photoresist and mask prevent a conclusive contact and thus worsen the quality of the imaging. Moreover, the close contact can cause scratching of the photoresist film on the substrate or the photomask itself.

Despite high attainable resolution this economically suitable procedure is used only rarely in the industrial manufacturing because of the above named disadvantages. Within the research area, which is not oriented toward maximum yield, this procedure enables a low-prized production of samples with structural sizes within the submicrometer range. For nanometer scale applications minimum structural widths of about 40 nm are obtained by direct isotropic etching of the photoresist film after developing [227].

Non-contact Exposure (Proximity)

With this procedure the disadvantage of the close contact between substrate and mask is eliminated by which the wafer is kept reproducibly at 20–30 μm away from the mask by means of defined spacers. Therefore few errors or contaminations occur both in the resist layer and at the mask.

The UV exposure delivers a shadow image of the mask in the photoresist. However, the resolution clearly decreases as a result of the proximity-distance; due to the diffraction effects at the chromium edges of the mask only structures with smallest dimensions down to about 2.5 μm are resolved. For the nanometer lithography these devices are completely unsuitable. In the industrial production the proximity exposure is also only rarely in use because of the insufficient resolution. An improvement of the resolution by advancement of the devices does not take place.

Projection Exposure

The resolution of the projection exposure procedure is determined by the light wavelength, the coherency degree of the light and the numeric aperture (NA) of the lenses. For the smallest resolvable distance a we get:

$$a = k_1 \frac{\lambda}{NA} \tag{7.1}$$

For the depth of focus (DOF) which should amount to at least ±1 μm because of the usual resist thickness in combination with surface irregularities and the focus position, holds:

$$DOF = \pm k_2 \frac{\lambda}{NA} \tag{7.2}$$

k_1 and k_2 are pre-factors which take into account both the entrance opening of the lenses and the coherency degree of the light, and the resolution criterion. Typical values for NA lie between 0.3 and 0.6; k_1 amounts to about 0.6, k_2 to about 0.5 for incoherent light.

From the equations a linear improvement of the resolution occurs with shrinking wavelength, but also corresponds to a linear decrease of the depth of focus. With $\lambda = 248$ nm, the typical used wavelength within the deep UV range (Deep UV, DUV), the depth of focus of today's devices is only insignificantly larger than the thickness of the photoresist. The minimum attainable line distance according to these equations amounts to about 250 nm with a depth of focus of about ±0.6 μm.

While the 1:1 contact lithography outweighs within the research area, in the industrial production devices for projection lithography are mainly used preferably as scanners for the exposure of the substrates. KrF laser or ArF laser serve as light sources: the used wavelength amounts correspondingly to 248 nm or 193 nm. The wafer scan procedure, the step and repeat exposure, and the step-scan procedure are used (Fig. 7.15).

The wafer scan procedure uses a lens system made of quartz glass for the 1:1 projection of the complete mask structures on the photoresist layer of the substrate. The exposure takes place via single over-scanning of the mask with a light beam expanding in one direction. In comparison to homogeneous illumination of the mask with 1:1 projection exposure the demands on the lens system in the scan

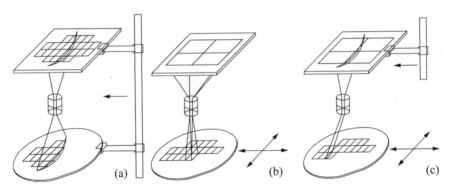

Fig. 7.15 Comparison of the exposure procedures: (a) wafer scan, (b) step and repeat, (c) step scan procedures

method are substantially smaller. Lens aberration can also be corrected more simply. The resolution limit of the wafer-scan-systems lies in the range of about 0.5 μm in line width depending upon source of light.

Due to variations in temperature during exposure and thermal processes during the substrate treatment, deviations in the adjustment accuracy can occur in the 1:1 projection exposure, from the center of the substrates to the boundary regions as a result of distortions. An adjustment of the mask fitting to all structures on the entire substrate is no longer possible so that the number of correctly processed elements reduces.

For this reason there has been a transition from complete exposure to step and repeat exposure in the mid eighties. Only a small reproducible fundamental unit is produced as mask. This is adjusted to the substrate and projected in the photoresist via a lens system. By repeated adjustment and exposure the complete structural imaging takes place on the substrate.

Transfer scales of 1:1, 4:1 and 5:1 are usual, whereby reduced projection exposure enables a better structure control of the patterns. Since the lens system must illuminate only a part of the substrate surface, it can be manufactured simpler and less expensive than in the case of complete exposure. However, its disadvantage is that it is time consuming for the repeated positioning and adjustment of the wafer transfer units to the mask.

The attainable resolution of these devices currently lies in the range of 150 nm line width, the adjustment accuracy is almost continuous over the entire substrate, deviations from chip to chip are so far negligible. Possible available particles within the mask area are image reduced, hence they partially fall below the resolution limit and are no longer imaged by the lens system.

In order to reduce the costs of the high-quality lenses as low as possible, reduction projection scanners are increasingly used. By simultaneous synchronal movement of the mask and the substrate with a fixed unit from light source and lens system large chip surfaces can also be exposed by over scanning with reduced lens diameter. Distortions by lens aberrations are simpler to compensate in these devices. The minimum structure size attainable with this method will be reduced presumably to about 100 nm or less in the next years.

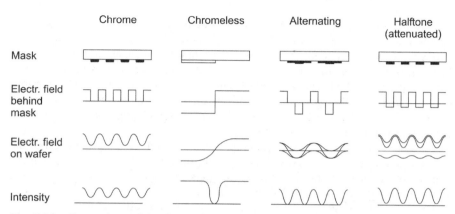

Fig. 7.16 Comparison of the distribution of intensity at the substrate surface for a chromium mask, the chromeless phase mask, an alternating chromium phase mask and the halftone phase mask

The necessary adjustment and overlay accuracies of the photolithography steps are achieved by means of interferometric position control and very exactly regulated processing temperature during the mask making and during the exposure. By further optimization of today's usual techniques the future requirements can be met in these dimensions.

The resolution of the optical lithography technique is limited by diffraction effects at the structure edges of the chromium layer on the mask. In order to get a more favorable distribution of intensity on the disk surface, increasing alternative mask designs are used. Absorbing phase masks can be used to replace the simple imaging chromium masks. They do not completely absorb an incident electromagnetic wave within the masked area, but only strongly absorb it and shift its phase by 180°. A more favorable distribution of intensity and thus a stronger contrast occurs on the substrate surface by means of interference.

Additional absorbers are partially produced on the mask which cannot be resolved by the used lens system any longer but effectuate an improvement of the structure transfer from the mask pattern into the photoresist by means of diffraction.

A further development is the chromiumless phase mask. By structuring of the mask material within the imaging area a phase shift of the electromagnetic wave by 180° is locally adjusted so that with a given irradiation wavelength a steeper transition occurs from exposed to imaged sub-area on the wafer surface.

The most favorable distribution of intensity for structure transfer is produced by the half-tone phase mask. With this design the absorbers reduce the incident electromagnetic wave up to a rest transmission, at the same time the light experiences a phase shift of 180°. A high contrast image of the mask information transferred into the photoresist results. The production of these masks is clearly more simple in comparison to the alternating chromium phase mask. However, their structure calculations are complex.

In order to achieve a further improvement of the resolution, sub resolution structures, hence samples with dimensions below the resolution of the applied optics are used for the correction of the distribution of intensity at the wafer surface. Thus, by diffraction or interference effects resolution improvements in corners, on points, and particularly with isolated lines can be obtained. The distribution of the sub resolution structures and the phase shifting elements in the mask must be calculated with efficient computers and be transferred precisely into the quartz mask in the dry etching technique.

Currently, the electron beam lithography is employed for the production of the required highly-resolved masks. Mechanically operating devices such as pattern generators are seldom used. Their resolutions are not sufficient for structure widths below 350 nm on the substrate.

Writing of the mask with line widths around 100 nm is time consuming. However the devices available today operate stably in the required time span. In prognoses, resolutions around 20 nm are asserted for mask writing with electron beams.

In order to increase the yield in the mask making, mask repair tools with lasers are available for subsequent exposure or for etching. These must be replaced by FIB (focused ion beam) systems during further reduction in the structure size since the focusing of the laser beam spot is no longer possible on dimensions in the nanometer range.

For the reduction of diffraction effects at the structure edges of the masks the light source in the projection exposers are developed further. Today, "off-axis" illumination is used in place of the point-like light source which was used as standard over decades. While with central illumination of the mask both the unbroken light beam and the −1 and 1 diffraction orders contribute to the imaging, the off-axis light source causes a suppression of a diffracted beam, e.g., the −1 order.

First improvements were obtained with circular light sources. More favorable results are achieved by the quadrupole or CQUEST II intensity distributions as light source. The latter consists of four symmetrically arranged light sources with a weak total surface superimposition as basic intensity (Fig. 7.17).

Substantial improvements in the resolution are possible by application of high-contrast resists or by multi-layer systems with thin radiation-sensitive surface films. Beside the already currently wide-spread anti-reflection layers as top or bottom coatings for sensitivity optimization and suppression of reflexes, changed resist systems such as CARL (chemically amplified resist lithography) [229] or

Fig. 7.17   Off-axis exposure for the reduction of diffraction effects, from left to the right: Standard, annular, quadrupole CQUEST I, quadrupole CQUEST II

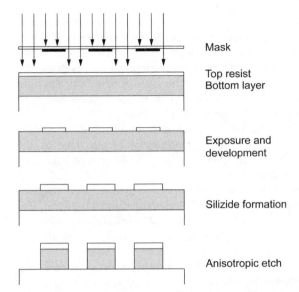

Fig. 7.18 Chemically amplified resist lithography (CARL)

TSI systems (top surface imaging) [230–232] are suitable for improving resolution.

The procedures use a sequence of layers from a thick masking resist which is covered with an extremely thin photo-sensitive layer in place of the usual resist. Only the thin layer is exposed to high-resolution via a mask; with this thin film, depth of focus and diffraction effects do not have significant negative effects. After developing, a very thin but highly resolved resist structure is present.

The produced structure is generally not suitable as an etching mask. It is firstly reinforced by a subsequent thermal or chemical treatment. Afterwards the structure is transferred by anisotropic dry etching, mostly in oxygen plasma into the masking bottom layer below. The masking layer then serves for the structure production in the active layers of the substrate. Applications of this resist can be found in microelectronics with line widths below 150 nm.

7.3.3 Perspectives for the Optical Lithography

Although the limitations of the optical lithography are predicted for years, these do not seem to be achieved yet. Line widths of 100 nm, possibly of 70 nm or even 50 nm, can presumably be transferred by optical lithography. It is doubtful whether a further resolution improvement up to 35 nm structure width is possible.

An increase of the resolution is aimed at by reduction of the wavelength down to 156 nm (F_2 Laser) and further down into the x-ray regime (EUV, extreme ultraviolet). However, new optics have to be developed for the projection lithography since quartz lenses age or lose transparency due to radiation stress (production of color centers, missing transparency of the materials for this wavelength).

Calcium fluoride lenses are used for the 156 nm radiation. Moreover, a transition to reflecting or mixed refracting/reflecting optics (catadioptrics) is discussed or is already used in the development labs. The entire path of rays from the light source to the photoresist must run in the vacuum or in an inert gas atmosphere since oxygen molecules lead to the absorption of the photons.

EUV lithography is seen as a continuation of the optical method in the context of further reduction of the wavelength to approximately 13 nm. Due to the wavelength within the x-ray regime, operation can be done only with reflecting optics which is currently in the development stage.

The structures which have to be transferred are produced by a reduced image of a reflecting mask in the photoresist applying wavelengths around 11–14 nm. Reflecting optics in the form of multilayer mirrors are used as optical elements.

With reference to today's level of knowledge multilayer systems from silicon-molybdenum films are suitable as mirrors with reflectivities about 70 %; this means a remaining intensity of maximum 8 % at the substrate surface for an optical system of 7 elements. The technological hurdles of this procedure exist essentially in the guarantee for surface quality of the optics over larger areas and the availability of efficient radiation sources in this wavelength range. Since the masks must also be produced as a reflecting element, a new development is required in this area. Bragg reflection can be used on a series of thin layers; the

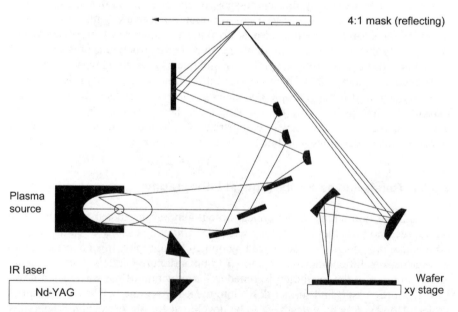

Fig. 7.19   Structure of a EUV step-scan exposer with plasma source and reflex mask, according to [228]

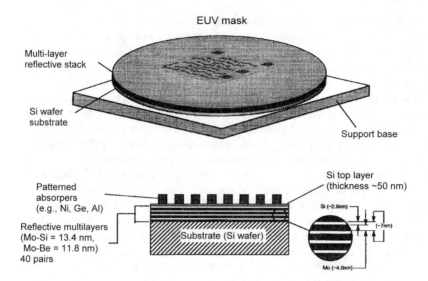

Fig. 7.20   Structure of a multilayer reflex mask for EUV exposure [235]

thicknesses of the respective layers must be controlled very exactly. An example of a mask is shown in Fig. 7.20.

At present, a laser-generated plasma is favored as radiation source with which pulsed laser light is focused on a collection of cooled xenon clusters. The xenon atoms are so strongly heated that characteristic radiation is emitted within the range of interest between 11 and 14 nm (Fig. 7.21).

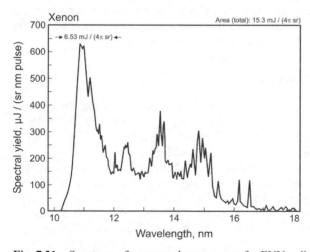

Fig. 7.21   Spectrum of a xenon plasma source for EUV radiation

For sufficiently short exposure times, radiating power within the range of approximately 20 W in a bandwidth of approximately 0.2 nm, given by the mirror optics, is required. For this purpose pulsed lasers are needed with a pulse duration of typically 10 ns and an average power output of some kilowatts. Such lasers are not yet available today; they still require further research for some years.

Because these sources are not available, the interest in gas discharge-based EUV radiation sources increases. With these sources the plasma emitted within the EUV area is produced by stored electrical energy in form of a pulsed discharge. In comparison with laser-generated plasmas, gas discharge plasmas offer the principal advantages of a more direct and thus more effective transformation of the electrical energy into light, a simple, more compact, and concomitantly low-priced setup, and a reduced debris problem.

Thus, at the Sandia National Laboratory [231] in California work is done on capillary discharge as alternative to the laser-produced plasma. The main problems emerge in achieving the necessary life time and achieving the necessary repetition rates or the required average radiating power.

In comparison with radiation sources examined so far by new electrode geometry, a new gas discharge radiation source [232] clearly promises higher operating life and repeating rates within the range of some kilohertz. Such repeating rates are necessary in order to ensure a sufficiently high average radiating power. Figure 7.21 shows an emission spectrum operated with xenon as discharge gas.

In comparison to the laser-produced plasma this radiation source is substantially more simple, more compact, and low-priced. The spectral characteristics are comparable with those of other radiation sources and fulfill the requirements of EUV lithography. The life time limiting erosion of the electrodes does not occur here, and a modification of the emission characteristics has not been notified. In common with negligible electrode erosion the debris problem can be neglected. With already currently attainable radiating power within the range of several 100 mW this source lies world-wide at the apex of gas discharge-based sources and in the range of the best laser-produced radiation sources. Compared with the other concepts this radiation source has a substantially high potential of fulfilling the requirements of the EUVL in a few years.

7.3.4 Electron Beam Lithography

In electron beam lithography, like in the case of direct writing of the photomasks, a finely focused computer controlled electron beam is scanned over the substrate coated with a special electron beam sensitive resist. The areas which are not to be exposed are blanked i.e., they are not illuminated with electrons.

For irradiation the semiconductor wafer coated with electron-sensitive resist must be transferred into the high vacuum of the system. There, scanning can take place line by line (line scan procedure) or in the vector scan procedure, whereby the latter manifest a higher throughput. Since not only chip for chip must be written but also each structure of each chip, the exposure procedure is time consuming. In order to minimize the writing time, the beam width in the point of focus can be continuously varied during writing (variable beam shape).

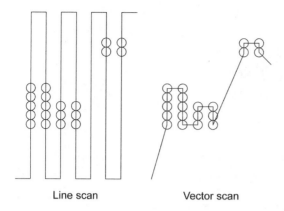

Line scan   Vector scan

Fig. 7.22   Comparison of the line scan and vector scan procedure

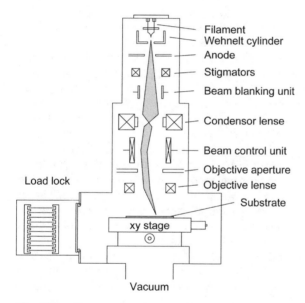

Filament
Wehnelt cylinder
Anode
Stigmators
Beam blanking unit
Condensor lense
Beam control unit
Objective aperture
Objective lense
Substrate

Load lock

xy stage

Vacuum

Fig. 7.23   Cross-sectional view of the electron-optical column of an electron-beam writer (according to [228])

Although the structural resolution during the electron beam exposure meets all requirements of future lithography techniques, this procedure is used mainly for mask manufacturing for the optical lithography because of the small throughput. It is only profitable in special cases for direct writing on substrates, e.g., for mask-programmed circuits with small number of elements.

The resolution of the electron beam lithography procedure of modern devices with finely focused beam is clearly smaller than 30 nm in line width; 5 nm structure widths are partly achieved. However, the writing time increases strongly with

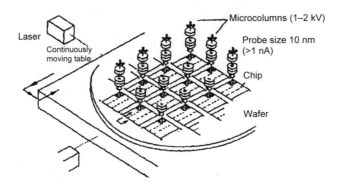

Fig. 7.24 Multi-column electron beam exposure [233]

the required resolution, so that for direct writing of very fine structures irradiation times of some hours/substrate are expected. Currently, the purchase cost of the new equipment is approximately 15,000,000 US$.

The electron beam exposure offers the possibility of exposing the individual layers of a chip with structure sizes down to 30 nm width fast and differently from wafer to wafer without taking the expensive route of mask manufacturing, particularly within the area of the specific applications in integrated circuits (ASIC). Thus, small chip numbers can also be manufactured relatively low-priced despite the high device costs.

In order to compensate the disadvantage of the long writing time for each wafer respective the small throughput, electron-beam writers with several independently controllable beams are currently being developed. Both multi-beam writers (multi beam) and writers with many micro electron columns (multi-column) [233] are in development labs. However, the independent alignment and the focusing of the individual beams are very expensive. Moreover, the simultaneous control of many electron beams requires an enormous data throughput.

Alternatively a technique of reducing electron beam exposure with an electron scattering mask is developed (SCALPEL, scattering with angular limitation projection electron beam lithography) [234]. The procedure uses a silicon nitride membrane transparent for electrons which is strengthened in the masking area with a metallic scattering layer. Electrons, which hit this scattering layer, are strongly deflected while electrons which hit the membrane only slightly modify their direction of propagation.

After focusing of all electrons, the aperture diaphragm fades out the strongly scattered electrons and only the particles with small diversion pass through the aperture and lead to exposure (cf. Fig. 7.25). The heating up of the mask is relatively small because the electrons are only scattered but not absorbed.

A resolution of 30 nm line width has been demonstrated already. The very small illuminated mask surfaces which limit the chip size are disadvantageous.

The mask for the SCALPEL technique consists of a silicon wafer which is coated as diaphragm with silicon nitride. The scatterers made of tungsten are va-

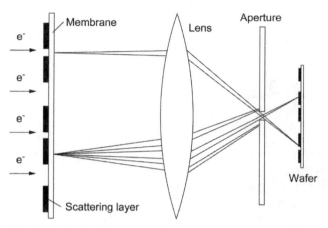

Fig. 7.25 SCALPEL method for electron beam exposure

por-deposited as scattering layer on this diaphragm. Subsequently, the structuring of the tungsten layer with conventional electron beam lithography and dry etching takes place.

In the last manufacture step of the mask the silicon under the nitride diaphragm must be removed; here the anisotropic wet-chemical etching with KOH or EDP solutions is appropriate. The resulting structure is presented in Fig. 7.26.

An alternative to SCALPEL is the PREVAIL procedure (projection reduction exposure with variable axis immersion lenses) [236], a further reducing electron projection technique. PREVAIL uses the so-called stencil masks, which consist of free standing structures in place of the diaphragm masks.

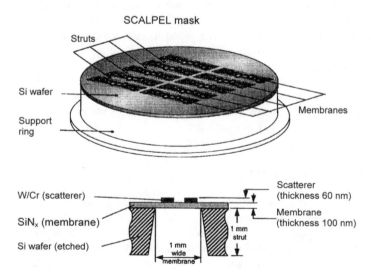

Fig. 7.26 Masks for the exposure according to the SCALPEL method [235]

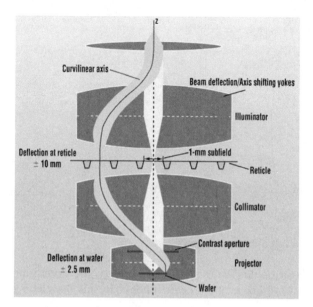

Fig. 7.27   Path of the electrons in the lens system of a PREVAIL exposure device [236]

The advantage of PREVAIL is the larger projection area by optimized control of the lens aberrations so that surfaces with chip dimensions can be imaged over a mask. The throughput of this procedure is correspondingly high. The disadvantage here is the heating up and thus the thermal expansion of the stencil mask due to the beam energy absorption.

For the first time, both PREVAIL and SCALPEL open the way for electron beam exposure with acceptable throughput. Due to the small imaging area of SCALPEL and due to the stencil mask of PREVAIL, the chances of both techniques for application in the production for nanostructured substrates are rather small in comparison with the progressive optical procedure with 156 nm wavelength or the EUV exposure.

Multiple beam electron systems or also the method of the exposure of the photoresist with tips of a scanning tunneling microscope for maximum resolved structures are still a far away from industrial application despite promising applications in research labs.

7.3.5   Ion Beam Lithography

Ion beam lithography is used on the one hand for projection exposure with masks but on the other hand it is also comparable to the electron beam lithography for direct exposure. In the case of the direct exposure with a finely focused ion beam, the higher particle mass in comparison to the electron mass causes a decrease of the required ion dose for resist exposure of about a factor of 10–100 in relation to the electron dose. Thus, the writing rate can clearly be increased. However, whether sufficiently high writing rates can be obtained for the exposure of 300 mm wafers in the manufacturing is unlikely so far.

Alternatively, ion beam lithography using an expanded beam with a diameter of approximately a square centimeter can be applied in the projection procedure. An example of a developed device is schematically shown in Fig. 7.28. Here, the mask must be laid out as stencil mask for image projection so that the ions between the absorbers can reach the photoresist layer on the substrate without dispersion.

In comparison to PREVAIL exposure, the stability of the projection masks is problematic using this procedure. Since we are dealing with a reduced projection exposure, the mask consists of an appropriately increased pattern. A thin silicon diaphragm, which is etched from a single-crystal silicon wafer, serves as mask material. The required mechanical stability of the mask can be achieved by means of a back-up ring at the edge of the disk. However, the thermal stress of the mask structure leads to an uncontrolled distortion of the structural pattern due to the absorption of the ions. This can be reduced by a small dose rate but the exposure time per substrate consequently grows. Moreover, double irradiation with supplementary masks is necessary for the exposure of special structures.

7.3.6   X-ray and Synchrotron Lithography

Because of the substantially smaller wavelength in comparison to the optical lithography, diffraction patterns at structure edges occur only with structural widths well below 100 nm when using x-rays. Therefore finer structures can be imaged with x-ray lithography than with optical procedures. The wavelength of approximately 1 nm promises a considerably higher resolution. However, due to the Fresnel diffraction and because of the generated photo electrons, limiting effects occur for the minimum attainable structure width, so that the limit of the resolution as a

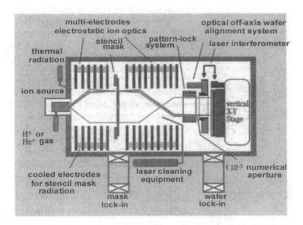

Fig. 7.28   Setup of a device for ion projection exposure with a stencil mask [237]

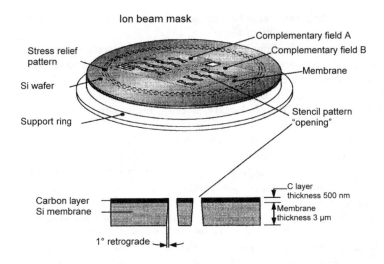

Fig. 7.29   Mask structure in the ion beam lithography [235]

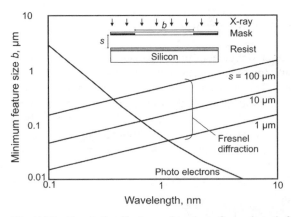

Fig. 7.30   Resolution limit as a function of wavelength for x-ray lithography

function of the projection distance lies in the order of approximately 70 nm (cf. Fig. 7.30).

X-ray lithography also operates according to the step and repeat procedure with a 1:1 mask, which is transferred as the shadow image in the proximity distance. Plasma sources (see EUV) or synchrotron radiation are used as x-ray sources.

X-ray lithography requires a mask with non-absorbing material instead of the usual quartz masks with a strength of about 3 mm and coated with chromium. Therefore, the substrate of the masked layer must have a low ordinal number (beryllium, silicon nitride or silicon carbide) and must be present in the form of a thin, mechanically stable foil (thickness of approx. 5–10 μm).

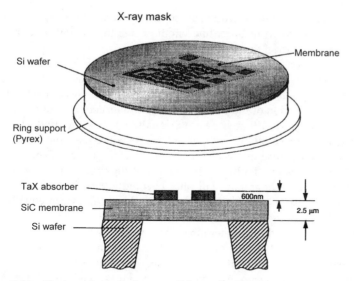

Fig. 7.31 Design of a mask for x-ray lithography [235]

Local masking cannot be implemented by chromium layers. Here, heavy elements such as gold, tantalum or tungsten are required for the absorption of radiation. An intensity ratio of 10:1 is achieved between the permeable and the impermeable mask areas.

The structuring of the masks for the x-ray lithography takes place with the help of the electron beam technique. The absorber layer is deposited galvanically on the thin carrier membrane, whereby the exposed resist mask in negative technique releases only the desired structures for coating.

Altogether the mask making is very complex and expensive, whereby the necessary size accuracy is not yet satisfactorily solved. Nevertheless x-ray steppers are already sporadically exploited in the industry.

X-ray lithography has been unsuccessful so far despite intensive research over more than two decades, because the optical lithography is substantially more simple to execute with the previous line widths. X-ray lithography may be a solution for structure widths between 70 and 40 nm. Presumably, finer structures cannot be imaged.

7.3.7 Evaluation and Future Prospects

The optical lithography could not be replaced so far in the industrial production because on the one hand, there is constant development of its resolution and on the other hand the current usual structural widths of down to 130 nm are also economical to execute.

In the following years the exposure wavelength will continue to decrease. Therefore the resolution of the devices will rise. With regard to multi-layer resist systems the attainable minimum structure size of this procedure will shrink down to 70 nm, possibly even down to 50 m line width. Thus, the optical lithography will dominate further at least in the next 5 to 7 years.

A foreseeable successor for the optical lithography is the EUV lithography. The reflecting optics enable highly-resolved reducing images of reflecting masks, the throughput will be comparable to that of the optical procedure. Nevertheless, it is crucial whether the application of the EUV technique counts economically.

Electron beam lithography will be used provisionally only for the production of masks. All efforts to increase the throughput of the procedure by projection techniques fail presumably because of the complex mask technique or because of the stability of the masks. Despite the higher resolution of the procedure an application of the electron beam lithography will only become interesting for structure sizes under 50 nm because of cost reasons. However, these values can equally be attained with the EUV lithography.

Ion beam lithography enables a higher throughput in comparison to the electron beam exposure but the stability of the masks is the application limiting factor. The purchase and operation cost as well as the missing availability of commercial systems for large area ion beam exposure do not create any place for this procedure in industrial production.

X-ray lithography will not be successful in production. Beside the complex radiation source the expensive mask making affects it negatively. High resolutions can be achieved only by a very small proximity distance; this simultaneously effectuates a small yield due to developing resist defects.

In summary, the optical lithography will be dominant for line widths of approximately 50 nm, afterwards the EUV lithography will be used. Alternatively the electron beam lithography with the SCALPEL technique is a successor with restrictions for the exposure. However, only few manufacturers will probably penetrate into this structure size regime.

7.4   Focused Ion Beams

7.4.1   Principle and Motivation

The implantation of high-energy particles in solid states has a fixed value in modern technology, which developed historically from different procedures. On the one hand it is well-known for a long time that accelerated particles with high surface doses (more than about 100 particle/surface atom) cause an erosion effect at the surface, which can be used for structuring. With an implantation surface dose of about 1 particle/surface atom amorphization of the solid state occurs which can be used, for instance, in wet chemical etching via the then strongly increased solubility. On the other hand there is a high interest in doping semiconductors within the dose range of about 1/100 particle/surface atom. This has been implemented technologically with diffusion methods of deposited dopant layers, which run at relatively high substrate temperatures and which smear the intentional doping steps. The ion implantation has totally replaced this diffusion technique, since the doses can be substantially controlled more exactly via beam current and exposure time, the depth distribution more defined and usually no substrate heating is necessary. Thus, ion implantation plays a central role in the modern semiconductor technology.

However, up to now ions are always implanted on large areas through open windows in the resist, i.e., defined laterally by the photolithography. The process steps which can be performed are numerous and critical to impurities: spinning of the resist, backing, mask positioning, exposing, developing, rinsing, drying as well as removal (stripping) of the resist after the implantation. Additionally, large sections of the implantation dose are wasted in the resist, which in the long run also costs acceleration time. Therefore, there is significant motivation to bring dopants masklessly into a semiconductor. Here the focused ion beam implantation (FIB) [238] is required: if it is possible to extract ions of desired dopants from a microscopic source they can be accelerated and focused in a particle-optical system, in order to intentionally dope a semiconductor with lateral resolution. The beam deflection and in/out switching occur with a computer which can select different ion types even by means of a mass filter, in order to manufacture for example, complementary doping profiles. In addition, this application with a small dose range can be extended to high doses: amorphizing and sputtering can equally be laterally resolved with FIB, which enable an analysis of cross sections on the wafer without having to break the wafer. Even conductive strips can be cut open and be joined in other places by FIB enhanced gas deposition, in order to correct photolithography layout errors in small series directly on individual devices. Likewise, a trim of devices is conceivable whereby substantially more influence can be gained on the functionality than, for example, with laser trimming.

In this section, the conceptual and practical criteria as well as the equipment of the FIB technology will be discussed. This section does not claim completeness. The author merely attempts to give as broad an impression as possible about this field.

7.4.2 Equipment

Production of an Ion Beam

The focused ion beam technology (FIB) is based on possible point-like ion sources, which are referred to as emitters. These can be operated cryogenically and then yield elements like hydrogen, helium, nitrogen, and oxygen which are only present in the gas phase at room temperature and pressures below one atmosphere. But, since the substrate to be implanted is typically grounded and hence the ion source must be of high-voltage, cryogenic sources are relatively complex in construction and operation. With the "supertip", however, a He-ion beam, which has excellent point source characteristics, could be delivered [239]. Nevertheless, the life time of this source is only a few hours and thus, still too short for technical application.

Heavier elements such as metals and all others which can bond with metal alloys are won as focused ion beam from the so-called sources of liquid-metal ion source (LMIS) [240]. In the simplest case, their filling consists of only a single chemical element and can be isotopically pure in exceptional cases. However, it generally concerns an alloy, which is usually eutectic for the purpose of small melting point and whose constituents are selected on the basis of two criteria: 1. The requirement of the type of element, and 2. the eutectic compatibility. The latter is relevant since the necessary elements must be present in realistic concentrations in the alloy, manifest comparable partial pressure for the working temperature for preservation of the concentrations and should also be well extractable.

Alloys which have been developed and tested by A. Melnikov in the author's laboratory are listed in Table 7.1.

The extraction of focused ion beams from LMIS takes place a few millimeters away from a high-melting container or filament, in or on which a drop of the alloy is present and held via capillary forces. An equally high-melting needle (usually W) rises from the drop, which manifests a point often sharpened by electrolytic etching and must be moistened by the alloy. The extraction aperture which has a diameter of a few millimeters (typically 3 mm) and is often negatively charged with a high voltage of 4–9 kV is a few millimeters away from the needle. A liquid metal cone (Taylor cone) is formed at the point by electrostatic forces, whose radius of curvature of a few nanometers lies substantially below that of the solid metal point. At this point, ions are formed and extracted by the electrostatic point effect (excessive local field), which is still favored for the heating of the emitter (which is necessary for the melting of the alloy).

In the simplest case, which covers approximately 95 % of applications nowadays, the LMIS is filled with gallium. This metal already melts at 29 °C and therefore requires practically no heating. Heating to about 600 °C for some 10 seconds is necessary for moistening and remoistening of the metal needle in intervals of some 10 to 100 hours.

The life span of a LMIS depends crucially on the steam pressure of the ingredients at the working temperature and on vacuum conditions: the lower the steam pressure and the vacuum pressure is, the higher is the life time. With well pumped systems a vacuum pressure of about 10^{-9} mbar can be established on the LMIS,

Table 7.1   Alloys for LMIS

Alloy	Crucible	Melting point, °C
$Dy_{69}Ni_{31}$	Mo	693
$Co_{67}Dy_{33}$		755
$Ho_{70}Ni_{30}$	Mo	720
$Fe_{36.7}Ho_{63.3}$		875
$Fe_{18}Pr_{82}$	Ta	667
$Mn_{10.5}Pb_{89.5}$	Al_2O_3	328
$B_{45}Ni_{45}Si_{10}$	Graphite	900
$Au_{70}Be_{15}Si_{15}$	Graphite	365
$Au_{68.8}Ge_{23.5}Dy_{7.7}$	Mo	327
$Au_{78.2}Si_{13.8}Dy_8$	Graphite	294
$Au_{61.8}Ge_{28.2}Mn_{10}$	Graphite	371

with which a Ga-LMIS achieves a life time of typically 10^3 h without constant heating with an emission current of 10^{-5} A.

Structure of a FIB Column and Complete System

FIB columns are almost exclusively available commercially, home-made structures are very rare (according to the author's knowledge, this is done only in Rossendorf and Cambridge). Commercial providers are JEOL, EIKO, SEIKO (Japan) as well as FEI (USA) and ORSAY (France). FIB devices resemble scanning electron microscope columns, but the high voltage is reverse biased and all deflection units are electrostatic but not magnetic. Thus, it is taken into account that a magnetic field would sort according to impulses. This causes (often inevitable) difficulties in double images and other focusing errors by isotope mixture of the ion source.

The structure of a FIB column is shown in Fig. 7.32a, the total structure with scanning electron microscope in Fig. 7.32b. Usually the ion source lies on the positive acceleration potential (from some 10 to some 100 kV) and the target (sample or wafer) is grounded. The focusing elements are the so-called single-lenses which are composed of disk packages that direct the ion beam via drillings of approximately 3 mm large situated as exactly as possible in the column axle in which the electron beam is carried. The disks lie alternately on a high voltage which corresponds to about half of the acceleration voltage and the earth potential, through which focusing-defocusing is effectuated. However, the total effect is focusing which can simply be realized by the curved lines of the electric field in the surroundings of the holes, where the field is strongly inhomogeneous. With a single lens, there are two ways for electrical switching.

Technically, the easiest way is to tap the focusing voltage from the positive accelerating voltage simply by a voltage divider which requires only one high voltage tank (which in general is like the column isolated by quenching gas via SF_6). The ions which pass through the lens are thus delayed. Therefore, this lens technique is known as "decelerating mode". Thus, the ions remain relatively long in the lens, whereby the focusing effect becomes stronger and a relatively small high voltage is sufficient. In the "accelerating mode", the lens package is occupied by a negatively high voltage. Thus, a second high voltage tank is required. However, this complex solution has the advantage that the attainable focus is about 10 % smaller. This small advantage is gained not only by the higher expenditure, but also by operation reliability: since the ions stop in the lens during a short time when in "accelerating mode", the focusing effect is smaller and the magnitude of the required high voltage is about 10 % higher. This could simply be managed if the leakage current dependency were not highly nonlinear. A single lens operates at a disk distance of a few millimeters and voltages of about 50 kV with field strengths of about $2 \cdot 10^5$ V/cm, which is close to the breakdown field strength (vacuum-pressure-dependent). Therefore, a small increase of the focusing voltage can lead to a strong rise in the leakage current of the lens and becomes a problem particularly by the associated timely focusing fluctuations starting from about 1 µA. Such leakage currents become relevant within the medium-high vacuum range and within the ultrahigh vacuum range. In high vacuum the rest gas works more easily as an extinguisher. They arise particularly from micro particles on polished plates (VA steel) of the single lens which causes electron field emission due to their small surface radii of curvature. Disassembling, postpolishing, cleaning, and assembling with adjustment are very complex. Healing *(conditioning)* by "nitridation" is more simple: in the stationary flow equilibrium of N_2 with a pressure of some 10^{-4} mbar, an increased focusing voltage (up to 100 % more) is applied, which leads to a bluish luminous plasma discharge and to nitrating the steel surface. These nitrides have an extraordinary dielectric stability and such "conditioning" is usually enough for operating the system for several months.

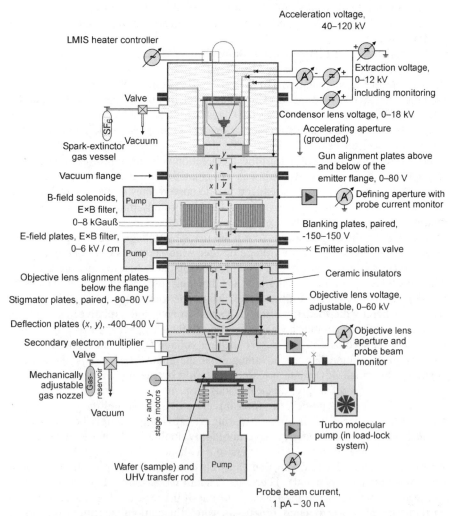

Fig. 7.32a   Schematic and functional setup of a FIB column

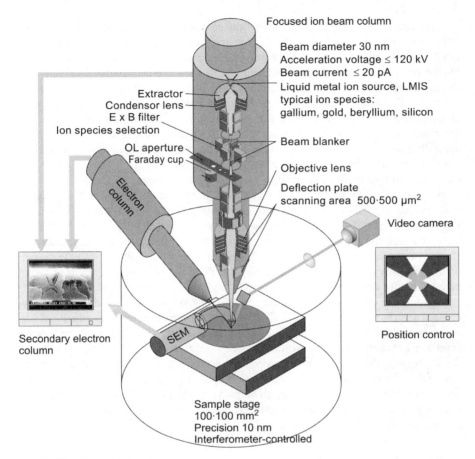

Fig. 7.32b   REM-FIB system

Beside the focusing elements electrostatic pair plates which are used for adjusting, dimming, and deflecting the FIB are needed. Static adjusting voltages are usually blocked like alternating voltage by RC filters in Hz range directly at the column near the pair plates in order to obtain high stability. Dimming and deflecting voltages can range from some 10 to some 100 V and must be available as a wide-band (MHz to GHz) in order to achieve high dose accuracies and writing rates.

Navigation and Joining of Write Fields (Stitching)

An unorganized search for details about maximum image field sizes of about 1 mm^2 particularly for large sample stages of 200 mm^2 and more is hopeless. As a result, strategies must be developed to discover certain places and a proper navigation is indispensable. If coordinates of the object are only roughly known, an optical navigation by means of a periscope optics and a simple CCD camera is very helpful. Thus, areas in the square centimeter range can be visualized.

Because of the writing field limitation of about 1 mm^2 larger structures can only be written if individual writing fields are precisely joined together *(stitching)*. On the one hand, the writing field must be as exact as possible for this purpose. On the other hand, the sample position must be measured substantially better than the beam diameter of the FIB, which is usually implemented by interferometric methods. The mechanical sample shifts are driven near the nominal position in approximately 0.5 µm steps (for the minimization of hysteresis always in one direction. In the opposite direction, about 10 µm are crossed and it moves back according to standard). The remaining difference between actual and nominal position is balanced by electrical correction of the deflecting systems, which can occur free of hysteresis in contrast to mechanical adjustment.

A substantial supplement of the stitching is the automatic mark recognition, which leads the FIB in the corners of the write field at right angle via etched or vapor-deposited cross thighs and records the secondary electron yield. As the ion beam crosses the edges of the cross thighs, the number of secondary electrons strongly rises and the actual position of the object relative to the coordinate system of the FIB can be determined automatically. Both the rotation and the translation orientation are considered and corrected.

In a similar automatic calibration mode of the writing field size and linearity, crosses are firstly sputtered for a short time in an unstructured and sacrificial object range which is set to an absolutely known position in the middle of the writing field via a sample stage controlled by laser interferometry. Then the FIB scans these crosses, determines the coordinates in the writing field by means of the above described automatic spot recognition and corrects the deflection factors and linearity parameters on the basis of a polynomial of fourth degree. Thus, the stitching is largely more exact since the edges of writing fields can then be implemented as straight line and orthogonal.

Image-Giving Procedures

Basically, FIB is exactly as image-giving as the scanning electron microscope: moreover, ion beams release secondary electrons from solid surfaces which have kinetic energy of only a few electron-volts and are easily sucked off by electrodes positively charged to about 10 kV and be detected fast and more sensitively in photomultipliers. Since the lateral straggling of the FIB is clearly smaller than that of electron beams, the secondary electrons originate almost exclusively from the impact area of the FIB focus and not from areas widened by the proximity effect like in the case of electron beams. In this regard, the image-giving FIB microscopy is still superior to the electron microscope having the same focus diameter.

Sample Transfer and Compatibility to Other Process Steps

FIB systems have been manufactured as ultrahigh vacuum (UHV) devices worldwide in some hundred copies. The cut-and-see high vacuum devices have conquered, above all, the industry in the semiconductor analysis with some 10,000 units since almost 10 years. The latter of course also analyze resist layers while organic samples are avoided in pure UHV-FIB devices for contamination reasons. This is also not necessary for the basic concept of FIB: focused ions permit maskless, resist-free, direct doping and sputtering of semiconductors which can then remain in the UHV during their entire processing. In many laboratories, particularly in Japan and the USA, MBE systems are connected with FIB systems via UHV vacuum tunnels since complete UHV processes can thus be executed. However, the author has good experiences with a UHV "suit case" concept, by which a CF100 UHV chamber (weight of approx. 40 kg) with window, personal ion getter pump, slide valve, transfer rod, and pump power supply with accumulator can be moved autonomously. During transport with a vehicle or train, the current can be supplied via 12 V dc or 220 V ac. Interruptions of about one hour in the circuit are uncritical for pressures below the 10^{-9} mbar range. This concept has the advantage of going back to many devices within or outside institutes or companies and enables a perfect oscillational decoupling of the FIB systems from the background, which is very difficult or even not feasible with vacuum tunnels. Since the UHV suit-case does not need to be ventilated over years, a strong baking and base pressure of 10^{-11} mbar are quite worthwhile. With this pressure the coverage rate of the remainder gas is about one monoposition/layer, which can be quite tolerated in most cases. In particularly critical applications such as MBE over growth after transfer, where the active layer lies directly on the transfer surface, this can be favorably covered with As in the case of III-V semiconductors. After the transfer, this protective layer is easily evaporated at temperatures of a few 100 °C and enables ultimate purities of, for instance, inverted HEMTs (high electron mobility transistor) which are grown over by MBE after the UHV transfer.

Thermal Annealing

After implantation thermal annealing must always be done in order to activate defect centers brought into the lattice, to anneal lattice defects, and generally to minimize long-term drifts in later operation. This can be done with different thermal procedures, whereby short process times are preferred due to smaller diffusion and of course lower costs. In most cases, complete thermal annealing is an excited process which can be described by the Boltzmann factor $e^{-E/(kT)}$ (E: excitation energy, k: Boltzmann constant, T: absolute temperature). To a good approximation, however, diffusion processes often run linearly in space and time whereby a short temperature pulse can anneal without releasing large and unwanted diffusions. This "rapid thermal annealing" (RTA) is executed in industrially compatible devices with halogen lamps (type 30 kW power for a 200 mm wafer), which achieve temperature ramps of about 300 K/s. The typical annealing temperature of 500–800 °C is held for about 10 s, cooling is done by radiation losses upon switching off the heating. Generally, RTA is performed in a mild inert gas atmosphere. If the material contains elements of high vapor pressure (like As in GaAs), it is usually sufficient to place fresh material of the same type directly on the surface of the processing material (face to face) in order to stabilize the partial pressure of the evaporating element.

7.4.3   Theory

Electrostatic Beam Deflection and Focusing

Magnetic inductions B always sort charged particles of the charge e and mass m according to impulses, since the Lorentz force $e\,v\,B$ is proportional to the impulse $m\,v$. In classical limes, the force $e\,E$ which is exerted on the charge particles by electric fields E, does not depend on the impulse whereby an electrostatic beam deflection and focusing influence all ions of a kind. Therefore, it is always of much advantage to conceive FIB systems exclusively electrostatically. By the high mass of the ions (relative to that of the electrons) their velocity is substantially smaller for comparable accelerating voltages, so that external magnetic perturbative fields play practically no role. In relation to the scanning electron microscopy, this must be rated as advantageous for the FIB.

Boundaries of the Focusing

Today's FIB systems achieve focus diameters from 100 nm down to about 8 nm. These values are favorably gained by sputtering holes in nm-thick Au layers and subsequent imaging. Of course the radial distribution of the FIB current is not ideally right angular, but approximately Gaussian in analog to optical beams. Here, there are also restrictions: only the first two orders of magnitude of the central current beam follow this distribution. Outside this domain the current beam drops almost exponentially and can therefore produce very unwanted "side doses". These are not of high importance in sputtering applications. However, they are quite disturbing during dopings with FIB.

Of course, the FIB, like the scanning electron microscopy, is not limited by diffraction effects like in the case of optical lithography: the appropriate de Broglie wavelengths in picometers are so small that they do not play a role. However, elastic and inelastic scattering processes for particle beams limit the resolution very much in the solid state: the lateral "straggling" of FIB lies in the order of magnitude of a tenth of the penetration depth. For electrons it is the penetration depth itself. Therefore, even if a very good focusing is achieved, they can be transferred in the solid state only to about this scale.

For the sake of simplicity the objective lens is usually operated only in the "decelerating mode". However, an ultimate solution is represented by the negatively biased "accelerating mode" objective lens.

The emission apex of the LMIS source has an expansion of only a few nanometers close of the point of the "taylor cones" formed due to the extraction voltage and thus is small enough to enable very high resolutions. However, this diameter cannot be maintained up to the sample, which is mainly because of chromatic lens aberrations of the objective lens. Single lenses focus particles of different impulses only if they are strictly of the same kind and have the same kinetic energies. The accelerating voltage can be kept constant, for instance, at 0.1 V which relatively corresponds to about 10^{-6}. The energy distribution of the extracted ions of a typical LMIS is however in the 10 eV range corresponding to a relative energy width of 10^{-4}. For the moment, this dispersion with the chromatic lens aberrations leads to a limitation of the focus to about 8 nm. In principle, the monochromatic character of the beam can be increased and thus the FIB focus could be further reduced with a high-resolution E×B filter.

The Wien Mass Filters and Its Resolution

The reasons specified in the section on electrostatic beam deflection and focusing sound mandatory for an electrostatic deflection. However, in order to be able to separate alloy sources ion types and charge states, an E×B mass filter is integrated in well developed FIB columns. There, the electric and magnetic fields are perpendicularly to each other and established on the axis of the columns. The magnetic field laterally deflects the ions according to the velocity-dependent Lorentz force. The oppositely oriented electric field brings the desired ion type back into the column axis where these ions reach the sample via the aperture. The other ions are absorbed at this aperture.

With this technique it is possible to extract p- and n-dopants for the doping implantation of semiconductors from well selected ion types of an alloy source, for example, Be and Si for GaAs, or B and P for Si. Thus, semiconductors can be directly doped without a mask and bipolar with only a single ion source by means of the FIB. Even if a ternary alloy is filled into the source whose third element is relatively heavy (for example Au, Ga, or the like), there is still a further ion source available controlled by direct electrical selection with which favorable sputtering can additionally be done.

The resolution of the filter is limited by the stability of the fields, their strengths and beam geometry (diameter and distance of the aperture). The fields can be sufficiently stabilized electronically so that this point is not critical. Permanent magnets for the B-field are particularly stable and elegant but must be removed when not required. This is why they are conveniently housed outside of the vacuum chamber. The typical attainable field strengths are $B \approx 1$ T and $E \approx 10^6$ V/m, the aperture distance about 10 cm, its diameters about 1 mm. Relative mass resolutions of about 10^{-2} are common so that the isotopes of gallium (^{69}Ga and ^{71}Ga), for example, can be separated well.

Thus, all elements available in LMIS (even pure isotopes) are practically implantable. This is quite relevant for special applications. However, in the case of alloy sources, spectral overlaps of different charge and mass states of different ingredients are possible. Therefore, the composition of an alloy LMIS should not be directed only on the desired ions and their vapor pressures but also on possible spectral interferences.

7.4.4   Applications

Single Ion Implantation

The current flow of a FIB beam should of course not be mistaken with that of an electron flow in a conductor. While very many free electrons exist in a metal due to the extremely large Fermi energy, but relatively only few can participate in the current flow, all ions in a FIB beam contribute *ballistically* to the current and therefore reach considerable velocities.

$$v = \sqrt{\frac{2E}{m}} = 440 \frac{\text{km}}{\text{s}} \sqrt{\frac{E\,(\text{kV})}{A}} \tag{7.3}$$

where v represents the velocity, E the kinetic energy, m the mass, and A the atomic weight of the ion. The left equal-to sign applies in SI units, the right one in practical units. For example, a velocity of $v = 526$ km/s which is commonly regarded for focused ions is obtained for a 100 keV Ga^+.

With a current beam I and elementary charge e per ion, of course I/e ions per second pass which hit the sample in timely intervals of e/I. The product of this time and the above velocity gives the average distance ℓ between the ions in the beam.

$$\ell = \sqrt{\frac{2E}{m}}\,\frac{e}{I} = \frac{7\,\text{cm}}{I\,(\text{pA})} \sqrt{\frac{E\,(\text{kV})}{A}} \tag{7.4}$$

This means, for example, that with $I = 1$ pA the average Ga ion distance amounts to 8.4 cm for 100 keV Ga^+. This pure macroscopic size suggests that with realistic beam current the ions have very large distances. Even with 1 nA the result is still $\ell = 84$ μm, which does not lead to considerable Coulomb repulsions or the like. Speaking figuratively, a FIB beam does not "flow" with currents even up to above microamperes like a connected water beam but only in small droplets whose distances are much larger than their radii (here the ions).

The above discussion illustrates that with a blanker, which manifest rise times of some nanoseconds and aperture distances in the 6 cm range (in systems of the Japanese company EIKO this is 5.75 cm), even light single ions are implantable through pulses of the blankers. Because of the quite high detection velocity of secondary electrons within the nanosecond range, even a feed back blanking is conceivable after impact of single ions. Thus, a substantially more defined doping of ultra small components can be implemented [241].

By the implantation of single defined impurities into unimpaired semiconductor areas, it is possible to study elementary electronic scattering processes and to examine transport equations such as the Boltzmann equation, which is normally applied only to statistical systems, also for this limiting case of single impurities.

Doping by FIB

Isolation Writing

FIB implantation like every high-energy implantation leads to lattice damages which localize free charge carriers. Therefore, these damages act isolating with which a local depletion and thus an isolation writing can be performed. Subsequent thermal annealing can reverse this isolation. However, for ion sorts which overcompensate a certain starting doping, of course areas remain depleted between the p and n regions. Therefore, a p-type line in a n-type area (e.g., a two-dimensional electron gas; 2DEG) works like two lateral anti-serial switched diodes. Thus, at room temperature it is also highly blocking in each polarity.

Lateral Field Effect

Isolation writing enables the creation of voltages of some volts between 2DEG regimes which lead to lateral electrical fields. These fields lie in the plane of the 2DEG and can change the lateral expansion by several micrometers or deplete areas of this expansion. If only some µm narrow channels are written, these are then typically depletable with -2 V via the neighboring so-called *in-plane gates* (IPG) [242, 243]. An enrichment is equally possible. However, it is limited to a relatively small effect of approximately 20 % of the channel conductivity since the additional charge carriers flow in much more disturbed areas. This effect can be used very elegantly in the so-called *velocity-modulated transistor* (VMT) [244]: in conventional field-effect transistors, the charge carrier density is changed to conductivity modulation which is a relatively slow process since charges must recombine or moved over large paths. However, because the free charge carriers in suitably written IPG transistors are shifted just a little in areas of essentially reduced mobility, the conductivity can be equally modulated. For this purpose, only the microscopic scattering process is exploited which is inherently fast.

With IPG transistors, transconductances of 100 µS and voltage amplifications of 100 can be achieved at room temperature, which is remarkable with regard to their total area below 1 µm^2. Writing velocities in meters per second are attainable with FIB columns which can lead to 10^6 components/s with lateral dimensions of the IPG transistors in the order of micrometers.

A substantial difference to conventional field-effect transistors is that with IPG transistor source, drain, and gate are written in one processing step with only two lines and therefore require no alignments. Beside the beam focus, the lithographic accuracy depends practically only on the resolution of the digital-to-analog converters which are operated with 16 or more bits on writing fields of 1 mm^2 dimensions. Thus, 15 nm resolution is not of special difficulty and is remarkable in view of the strong technological efforts in the UV optical lithography.

Positive Writing

While with isolation or also "negative" writing all written paths work isolating, a mode is also possible, in which the implanted areas become conductive by doping with suitable defect centers. With homogeneous semiconductors this of course leads to dispersion at impurities, which is accepted in today's Si technology. In heterostructures, however, *band gap engineering* can then be applied as well: because an undoped (empty) heterostructure is grown, for instance, by MBE, a higher band gap layer (e.g., $Al_{0.3}Ga_{0.7}As$) can be implanted by FIB into a near-surface layer so that the incorporated charge carriers are transferred into a deeper layer of smaller band gap (e.g., GaAs), where they have a higher mobility due to the absent defect centers scattering.

The production of IPG transistors is also possible with positive writing. The direct p-n junction or intrinsic areas are then used for isolation. The enrichment is substantially higher than with isolation writing since the carriers are shifted into unimpaired semiconductor areas. This favors the production of "normally-off" transistors which are essential for inverters.

A positive writing mode is also possible with regard to micromechanical structures. A free standing GaAs bridge with a length of 15 μm, a width of 500 nm, and a thickness of some 30 nm is shown in Fig. 7.33 [245]. It is written together with the 2·2 μm^2 square at its ends using 30 keV Ga$^+$ and a dose of 10^{16} cm^{-2}, whereby these areas are amorphized. Afterwards the GaAs that is not amorphized with FIB is removed up to a depth of about 3 μm using a selective potassium citrate etch for crystalline GaAs. This etch does not completely work isotropic as can be seen in the under-etched facet-like squares.

Fig. 7.33   Amorphized, free standing GaAs bridge with 15 μm length, etched in potassium citrate solution

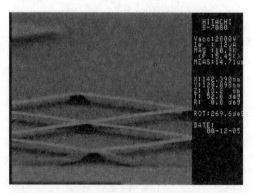

Fig. 7.34   Part of a free carrying lattice with a lattice constant of 10 μm manufactured like the bridge in Fig. 7.33. Lattice bars, which are only one-sidedly kept, are seen in the foreground and therefore by drying they are pressed on the substrate by adhesive forces.

A further positive process can be implemented with a gas inlet as described in the section on gas-supported sputtering: for this purpose, feed gases are let into the surrounding of the FIB point of impact which carry metal atoms on the one hand and remain volatile after the "cracking" on the other hand. An example is tungsten hexacarbonyl, $W(CO)_6$, which is present as a white powder under standard conditions. By heating up to 40–80 °C the material evaporates and can be let into the working chamber of the FIB as gas whereby the gas flow is well adjustable via the temperature. However, the whole length of the tube must be heated otherwise the danger of blockage exists. By supplying this gas and simultaneous FIB writing the molecules close of the surface are cracked and tungsten can be deposited metallically. Thus, a "post-wiring" also succeeds by low impedance, directly written conductive strips and together with the "cut-and-see" method, switching circuits can be corrected individually on the wafer.

Complementary Electronics by FIB

The possibility of having both types of doping available in one and the same FIB source opens a complementary doping technology. p and n-dopants can be selected simply by electrical switching and implanted via controlled software. This is mainly possible in the "positive" writing mode and enables a complementary semiconductor technology originating of only one process step whose throughput is greatly limited by the sequential writing method.

Newer developments in the formation of FIB systems use the purely electrostatic deflection and focusing in order to manufacture microcolumns with modern semiconductor technology [246]. Whole arrays of microcolumns could then write in parallel, thus allowing the possible use of such advanced devices in certain production areas.

Resist Lithography by FIB

Breaking Open Organic Resists

The present microelectronics is carried out exclusively by the optical lithography which opens windows in resist layers in order to locally facilitate etching procedures, large area implantations, metallizations, and oxidations. To a certain extent the optical lithography is completed by the electron beam lithography with which resolution-critical details are written partly in "mix and match" technique. Usually the bondings in the resist are broken open which is subsequently removed during the development process. This process is also possible with focused ions because ions have a substantially more aggressive impact on materials and their chemical bondings. A possible implantation contamination below the resist may be a limiting factor.

Cross-Linking Organic Resists

Cross-linking of open organic resists should be regarded as a complementary technique to their breaking. It is also possible to make resists more insoluble against developers by means of FIB, i.e., by cross-linking. Figure 7.35 shows lines of 36 nm width in intervals of 200 nm which have been prepared in this way.

FIB Sputtering

Cut and See

The possibility of sputtering and visualizing the sputtered object with the same particle beam has opened and still opens unforeseen dimensions in the micro and nanoanalysis. Visualization can be kept very less damaging by an efficient image storage and processing. Besides, the secondary electron yield by FIB depends more strongly on the considered material than with electron beams, so that the material contrast is clearly better in both semiconducting and metallic samples. Thus, domains and areas of strong doping gradients can surprisingly clearly be represented.

However, the main impact is the possibility of a morphological modification by FIB sputtering: components in integrated circuits "on wafer" can be properly dissected without dividing the chip (die) or by only dividing the wafer.

Structures in Diamond

While diffusion processes and cutting processing partly depend dramatically on the crystal structure and on the hardness of the material, implantation, and sputtering can be executed practically on every material, even on a solid state as extreme as diamond. This material has substantial advantages like high dielectric constant, high heat conductivity, high transparency (also in the visible range), high band gap, lowest leakage currents and highest hardness. Beside many other FIB applications the following two technologies are given as examples: 1. FIB direct writing of buried graphite conductive strips, 2. FIB direct sputtering of diffraction structures for integrated optics. The deceleration of the ions also takes place in a certain depth by the FIB penetration depth of some 10 nm. Within this range the largest part of the kinetic energy of the beams is transferred into lattice deformations. In diamond, graphite which is almost metallically conductive is formed while the surrounding diamond remains highly-isolating. The transition between

Fig. 7.35   Transversal cross-linked lines by FIB (width 36 nm, distances 200 nm) in photoresist on GaAs

graphite and diamond can still be intensified by RTA. Since the graphite formed in this way has a smaller density but is enclosed all about in the crystalline lattice, it is under high pressure that stabilizes the long-time behavior of the otherwise very fragile graphite. In this way buried conductive strips of smallest dimensions (for instance, in the FIB focus diameter) can be manufactured which can connect, for example, devices in diamond. The second exemplary area of application of FIB in diamond lies in the production of diffraction and interference patterns. By direct sputtering or wet-chemical etching after near-surface amorphization, some samples can be prepared on the surface of diamond, which could be used for instance in Fresnel lenses, photonic crystals or holographic structures. Such structures have the substantial advantage of a radiation hard, resistive, mechanically stable and dielectrically effective modulation in the optical regime.

Preparation for Succeeding Microscopy

The direct sputtering without a mask is also very effective for preparing cross sections for a succeeding microscopy, like already described in the "cut-and-see" procedure. Additionally the interesting possibility of preparing a very thin area that can be transparent to keV electrons by sputtering on both sides around a narrow area exists. The transmission electron microscopy (TEM) can often be operated at such narrow, often less than 100nm broad bars. Since relatively large areas must be removed for such a preparation in order to irradiate the bar parallel to the surface with the TEM, it can be appropriate, beforehand, to use wet chemical etching to deeply etch large sample areas via optical lithography and then to refine the bars by means of FIB.

Gas-Supported Sputtering

Purely physical FIB sputtering manifests two disadvantages: first of all the sputtering rate is limited and secondly a disturbing side dose often becomes apparent. Both disadvantages can be reduced if a gas which increases the sputtering rate is injected into the vacuum chamber near the FIB point of impact. For example, H_2O increases the sputtering rate of organic substances. Iodine gas or XeF_2 is often used in crystalline substances. The concept of the function is that these gases are divided or activated chemically by FIB and then the solid surface can be better etched. The fragments or reaction products are volatile to a large extent and are sucked off by the evacuated system. During this process, the operating pressure increases to a typical value of 10^{-5} mbar which can be very well mastered by turbo pumps. Crucial for an efficient and economical gas inlet are the diameter, the length, and the distance of the inlet nozzle from the point of impact of the FIB. Diameter and distance can be reduced to approximately 100 μm, the length of the inlet nozzle amounts to some centimeters. A thick pipe (inner diameter of about 5 mm) into which a refilling cartridge for the gas inlet can be slid is practically installed up to the external wall of the working chamber. The cartridge can be changed within some minutes via the thin inlet nozzle without breaking the vacuum of the working chamber, which rises to about 10^{-4} mbar in pressure through the defined leakage.

7.4.5   Evaluation and Future Prospects

In comparison with MBE and scanning electron microscopy, the FIB technology presented here is still in the preliminary stage from developmental point of view. With only a few hundred UHV research devices world-wide and hardly more specialists one could have the impression that the critical mass for enormous developmental swing is not yet completely achieved. However, a very wide market already appears for "cut and see" applications on the part of semiconductor analysis, which alone is evident of the fact that every considerable semiconductor manufacturer already operates at least one FIB high vacuum workstation and magnetic writing and reading heads for high-density hard disks, for instance, are mechanically retrimmed in the quantity production by FIB. The focus diameter of FIB is constantly improved from year to year, whereby a saturation is already recorded for a long time for electron microscopes. The basis for the preferred electrostatic deflection and focusing of FIB are excellently suitable for the miniaturization with methods of the modern semiconductor technology, whereby "multi-FIB" systems also appear achievable. These can significant improve the throughput of FIB lithography equipment by which this application also has a high potential future.

7.5   Nanoimprinting

Nanoimprinting, a technique which is surprisingly simple compared to the methods presented within the preceding paragraphs, has gained broad interest recently. In research it has already become an important tool for the definition of nanometer and sub-micrometer pattern and is impressive because of its ease of implementation and low cost level.

7.5.1   What is Nanoimprinting?

Generally, nanoimprinting stands for a number of methods where definition of lateral pattern of a surface layer is mechanically performed. The most important among these are embossing, printing, and molding. These techniques are characterized by the fact that a template (master, stamp) carrying the envisaged nanopattern is replicated on a thin surface layer on the substrate.

Some of these techniques are well known for patterning in the micrometer and sub-millimeter range typical for micromechanical devices and MEMS (micro electro mechanical systems) [247, 248]. The novel feature of nanoimprint is its application for nanometer patterning as well as its use as a lithography technique. The latter means patterning of a thin layer, usually an organic material (polymer, resist) on top of a substrate. This patterned layer serves as a mask or enables mask definition for a subsequent etching step (see Sect. 7.2) where the pattern is transferred to the substrate itself. Therefore in nanoimprint the template plays the role of the photomask in a conventional lithography process (see Sect. 7.3).

A detailed discussion of nanoimprint techniques focusing on lithography applications in given in [249, 250].

Nanoimprint Lithography (Hot Embossing Lithography)

Process. This technique has been proposed first by Chou [251, 252] and the term *nanoimprint lithography* (NIL) has become a generic term for all mechanical nano-patterning concepts. The process itself is a hot embossing process (*hot embossing lithography,* HEL). Its principle is explained in Fig. 7.36. A substrate, in most cases a silicon wafer, covered with a thin layer of thermoplastic polymer is heated up together with the template (stamp). At a temperature above the glass transition (glass transition temperature T_g) of the polymer, where its viscosity is sufficiently low to conform to the template, substrate and stamp are brought into contact and pressure is applied. As soon as the polymer has filled the stamp relief the stack is cooled to below the polymer's T_g and stamp and sample are separated, leaving the inverted pattern of the stamp frozen into the polymer layer as a thickness contrast. Finally the polymer remaining in the gaps, the residual layer, is removed in an anisotropic dry etch step, thus completing the lithography process. The so patterned polymer layer is now ready for use as an etch mask for pattern transfer to the substrate or as a mask for lift-off. For lift-off the patterned polymer layer is evaporated with metal. Due to the inhibited coverage of steep walls in an evaporation process the polymer can be dissolved, flooding away the metal on top of it. The remaining metal corresponds to the stamp pattern and may be directly used as an electrical wiring. Alternatively it may again serve as an etching mask for patterning of the substrate. Due to the superior mask selectivity of metals compared to a polymer this is the preferred technique for pattern transfer in the several nanometer range.

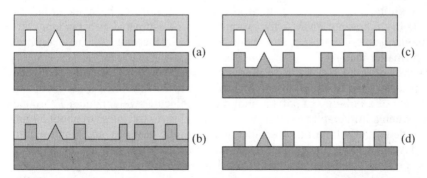

Fig. 7.36 Principle of hot embossing lithography (HEL). Patterned stamp (template) and sample (substrate with spincoated polymer layer) are heated to process temperature (a) and brought into contact. As soon as the polymer has conformed to the stamp relief under pressure (b) the stack is cooled down. Separation of stamp and sample (c) is done below the glass temperature T_g, where the thickness contrast is frozen in the polymer layer. The residual layer remaining is removed in an anisotropic dry etch process (d) for opening of the mask windows.

Characteristics. A number of resists used for classical photolithography (see Sect. 7.3) is likewise applicable as a thermoplastic polymer for nanoimprinting. Most groups use polymethylmethacrylate (PMMA), a polymer available commercially with a broad range of chain lengths (mean molecular weight <20 to 1000 kg/mol). Polymer size is an interesting parameter for nanoimprint, since at temperatures above T_g the viscosity and thus the ability to flow and to conform to the stamp strongly depends on the polymer's molecular weight [253]. High molecular weight polymers require higher temperature differences with respect to their T_g than low molecular weight polymers, when similar imprint behavior is envisaged. In addition to standard polymers like photoresists or PMMA, specific materials have been developed for nanoimprint [254] and are commercially available [255].

Stamps for nanoimprint made from silicon or from silicon with a patterned oxide or polysilicon layer on top are fabricated by conventional Si patterning technology. Stamps with nanometer patterns require complex lithography techniques and represent a major cost factor when large patterned areas are involved. For stamps with simple line patterns alternative techniques like holographic lithography or sidewall lithography (spacer technique) [256] may be applied for stamp fabrication. In addition, low cost working stamps have successfully been prepared by hot embossing into a curable polymer layer on Si [257].

It is inherent to the mechanical process that stamps with small patterns and in particular with periodic patterns are easily replicated. Stamps with larger and isolated patterns require higher pressure and temperature for successful replication, as transport of the polymer within larger lateral distances is required [258, 259]. A simplified hydrodynamic analysis of this squeezed flow [260] reveals, that the larger the stamp patterns and the thinner the polymer layer thickness the higher the force necessary for imprint [250, 261, 262]. As a consequence, nanoimprint is less suitable for definition of large patterns in thin layers due to these natural mechanical limitations. A lower pattern size limit has not yet been identified. A molecule of 20 kg/mol PMMA for example, can be estimated as a random clew of some nanometers diameter, and chain segments of the molecule may conform to much smaller features.

Pressures of up to 100 bar and temperatures of up to 200 °C are required when lateral pattern sizes of up to about 100 μm have to be imprinted in 200–400 nm thick layers of a polymer of 350 kg/mol mean molecular weight and a T_g around 100 °C, when a lithography process with a residual layer thickness of about 50 nm is envisaged [259]. Imprint of smaller and periodic patterns and imprint into thicker layers or lower molecular weight materials are successful at considerably reduced pressures and temperatures. Like discussed, due to the fact that the adequate processing parameters strongly depend on the stamp layout, universally valid data cannot be specified.

Nanoimprint is a parallel process, where the whole stamp area is replicated within one embossing step. Stamp areas of 4″ diameter and more have been tested, where anti-sticking layers on the stamp improve large area imprint quality when densely patterned stamps are involved. Often self-organized layers of fluorinated siloxanes are utilized [263] as anti-sticking layers, which can be applied from the

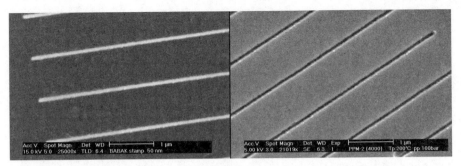

Fig. 7.37 Comparison of stamp (left) and imprinted polymer layer (right) in case of 50 nm wide lines

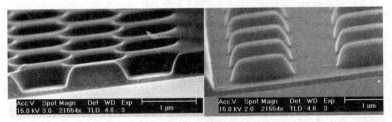

Fig. 7.38 Imprinted positive and negative dot patterns of about 1 μm diameter. To meet the requirements of a lithography process, the residual layer thickness of the polymer, as demonstrated in the SEM micrographs, is small compared to the overall thickness contrast.

liquid phase or the gas phase. The latter is preferred in the case of fine patterns as it overcomes wetting problems in narrow trenches. Overall process times lie in the range of minutes due to the heating/cooling cycle and also due to the fact, that the polymer needs time to flow. As a consequence of the above discussion, process time strongly depends on the required maximum temperature (and therefore on the polymer's T_g and its molecular weight) and on stamp layout (pattern size and pattern density). Investment costs are low, most research groups use a simple hydraulic press or a clamp devices in an oven. Application of nanoimprint for multilevel lithography is still scarce, where external alignment before mounting in the imprint system is the method of choice.

Results. Quite early, hot embossing lithography demonstrated the ability to prepare patterns of a size of 6 nm [252]. Our own work [259, 262] has shown that the method is able to reproduce a broad range of pattern sizes (see Figs. 7.37–7.40) and that it is also well suited for replication of three-dimensional patterns (Fig. 7.41) [264]. Imprint of 4″ diameter fully patterned stamps is under investigation [265, 266].

Mold-Assisted Lithography

Process. Mold assisted lithography (MAL) was introduced by Haisma [267]. A substrate is covered with a layer of a low viscosity UV curable monomer or oli-

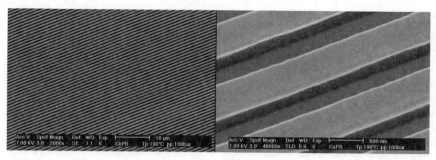

Fig. 7.39   Imprinted field (5·5 mm^2) of 400 nm lines (large area view and close up)

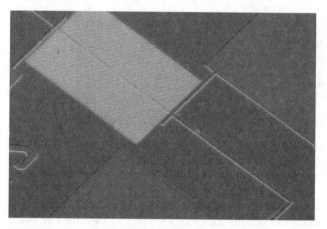

Fig. 7.40   Large area imprint of differently sized patterns: field of interdigitated lines of 400 nm width (top left of micrograph) and 100 μm square pad structures (top right and bottom of micrograph) located side by side. The imprint is done with a fully patterned stamp of 2·2 cm^2.

gomer which conforms to a mold under vacuum pressure. UV flood exposure through the mold crosslinks the patterned layer and fixes the surface pattern. Again a residual polymer layer has to be removed in order to finalize the mask.

Characteristics. The process relies on a UV-transparent mold, e.g., made from quartz, enabling optical alignment with respect to previous lithography steps. Because of the chemical cross-linking reactions taking place during curing an anti-sticking layer is inevitable for later separation of mold and substrate. A substrate adhesion layer may improve the spin-coating process of the monomer [267]. The procedure may be performed in parallel in a commercial contact printer for optical lithography when local non-uniformities of the residual layer are accepted [267].

At present the most promising procedure is a stepping one known as *step and flash imprint lithography* (SFIL) [268]. Here, a mold of several square centimeters is used and the monomer is directly applied locally between mold and substrate via capillary forces. Then the gap is closed to reduce the residual layer thickness and the monomer is cured within the mold region. Since only small forces are applicable optimization of the curable material with respect to a low viscosity is required, which, together with the restricted range of capillary effect limits the size of the usable molds.

In order to avoid the sticking problems reported in [267] in the case of molds with high aspect ratio, SFIL relies on low aspect ratio stamps and replicates the pattern on thin layers of Si-rich polymer compounds, which, after curing, are used for dry development of a thick underlying transfer layer. This transfer layer represents the final mask and provides the aspect ratios required for pattern transfer into the substrate in a subsequent process. Moreover, the thick transfer layer may serve as a planarizing layer when patterning over topography is envisaged.

Results. First publications on mold assisted lithography have already demonstrated arrays of dots with 35 nm diameter and about 100 nm pitch [267], as well as 70 nm wide lines transferred to the substrate with an aspect ratio of 3:1. SFIL has shown 60 nm patterns with aspect ratio 1:1, resulting in a mask with an aspect ratio of more than 6:1 after dry development of the transfer layer [269].

Microcontact Printing

Process. A third method analogous to a classical printing technique has been introduced by Whitesides as *microcontact printing* (µCP) [270, 271]. A template is replicated by molding on polydimethylsiloxane (PDMS), a thermally curable elastomer. This elastomeric stamp is inked with an alkanethiol. When in contact with gold (e.g., a thin sputtered Au layer on Si) the thiol is transferred from the elevated stamp structures to the substrate, forming a self-assembling monolayer (SAM) due to parallel orientation of the linear thiol molecules almost upright to the substrate surface [271]. The resulting patterned SAM layer has a thickness

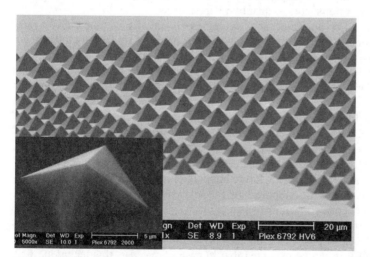

Fig. 7.41 The sharp edges of the imprinted three-dimensional pyramids demonstrate the pattern transfer fidelity of a hot embossing process.

corresponding to the chain length of the thiol molecule. It masks Au during a wet etch and serves as a mask for pattern transfer to the substrate.

Characteristics. Thiols result in low defect SAMs on Au and are well suited to mask the Au in a wet etch step due to their chemistry. SAMs may also be formed on oxide [272], but generally in reduced quality. Since Au is a deep trap in Si, µCP will mainly be applied for patterning problems beyond microelectronics for instance, for biological or chemical applications. Moreover, such applications benefit from the fact that preparation of the elastomeric stamp may be done from a patterned polymer, thus requiring only lithography. Therefore stamp fabrication is simple compared to the first two techniques where an additional etching step is needed.

Furthermore, µCP is easily performed on curved substrates and is robust to potential surface roughness, as long as the elastomeric stamp is flexible enough for their compensation.

Results. In its basic concept with a rolling procedure, µCP has been able to show patterns down to 300 nm over an area of 50 cm^2 [273]. Detailed research at IBM has optimized the stability of the elastomeric stamp significantly by developing a hybrid multilayer concept. With such stamps patterns down to 100 nm have been printed reproducibly over 25 cm^2 with relative adjustment accuracies of 1 µm [274].

7.5.2   Evaluation and Future Prospects

Due to the relatively low experimental and financial expenditure compared with the excellent results obtained in laboratory experiments the interest in nanoimprint techniques is still growing for research. Moreover, the number of research groups booms making use of it for preparation of specific nanopatterned devices ranging from single electron transistors to quantum wires.

Hot embossing. For hot embossing equipment is already commercially available [275], promising reproducible processing for areas of up to 6 inch diameter [265, 266, 276]. Adjustment is done externally (optical alignment, clamping and transport to the imprint system) so far, and applications concentrate mainly on patterning problems with only one lithography level (patterned media for storage, sensors, nanofluidics, nano- and microsystems, biochemical systems). Based on production tools for microelectronics a step and stamp procedure is investigated [277] and a combination with conventional optical lithography (mix and match) [278, 279] is tested to overcome the limits discussed when larger pattern sizes are involved. Present trends try to lower the processing temperatures and pressures. The potential of hot embossing for future application in production will depend on the progress achieved in development of flexible equipment, process technology and alignment concepts.

UV molding and SFIL. Development of production equipment and process technology for SFIL is pushed in intense cooperation between research institutes and industry [269]. It can be expected that in case of success, this technique will be applied to multi-step lithography and thus complex device production processes, despite of its limitation to moderate areas. A first automated stepper for 200 mm wafer diameter is under test [269].

Microcontact printing. The detailed development at IBM has demonstrated the potential of μCP for reproducible pattern definition down to 100 nm [274]. Due to the chemical masking nature of the SAM pattern the technique is rather suitable for patterning problems beyond microelectronics, particularly for biological and chemical applications where a number of replication and patterning techniques has established so-called soft lithography, possibly by avoiding the instrumentation of classical silicon technology [280].

7.6   Atomic Force Microscopy

7.6.1   Description of the Procedure and Results

The scanning tunneling microscope (STM) or atomic force microscope (AFM) has already been discussed in the section on the measurement of surface roughness. In this section, emphasis is put on its ability to pattern structures rather than to use it as a microscope. It was discovered that the STM is capable of shifting individual atoms with the tip of the microscope [281]. The authors operated with a highly pure (110) nickel surface being covered with single Xe atoms. Firstly, the tunneling current of the tip is fixed approximately at 1 nA, a value that is also adjusted for imaging. When approaching a Xe atom (Fig. 7.42a), the current increases to some 10 nA with fixed bias by lowering the tip. Thus, the atom is pulled onto the tip (Fig. 7.42b). Next, the tip is moved laterally to a new position with a velocity of 0.4 nm/s (Fig. 7.42c). There, the Xe atom is released (i.e., the current decreases) as the tip is retracted (Fig. 7.42d).

The result of this manipulation is depicted in Fig. 7.43. The authors report that it takes 24 hours to form the three letters I, B, and M.

7.6.2   Evaluation and Future Prospects

The work quoted above raised hopes in the area of nanoelectronics since one expected to break the limits of lithography. With the above-named technique, it

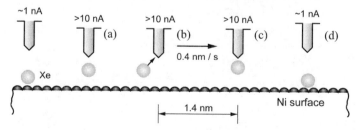

Fig. 7.42   Moving Xe atoms on a Ni surface with the STM

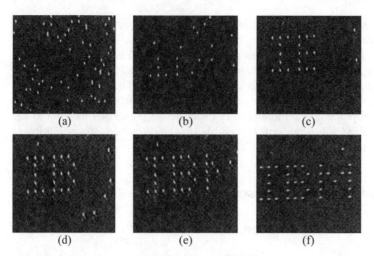

Fig. 7.43   Xe atoms are moved to form a pattern

should be possible, for instance, to manufacture metallization lines no longer by etching lithographic samples but by positioning individual metal atoms. With regard to the time needed to perform such a manipulation, substantial improvements will be necessary for a technologically useful application. A further application could be molecule synthesis where single atoms are brought into close contact.

7.7   Near-Field Optics

7.7.1   Description of the Method and Results

The basic idea of near-field optics is presented in Fig. 7.44. Light is directed at a sample S through an aperture A. The aperture can have a small diameter in relation to the wavelength λ of the light. Values down to $\lambda/50$ are possible. The sample must be placed within the near-field region, i.e., in the sub-wavelength range,

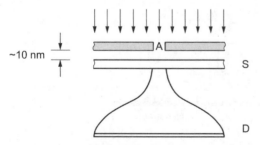

Fig. 7.44   Schematic setup of near-field optics [282]

in the proximity of the aperture. The light transmitted by the sample is expanded on the far-range and detected with a detector D. Please note that the local resolution is not restricted by the Abbe diffraction condition $\lambda/2$ anymore, but by the accuracy of the lateral positioning and the size of the aperture.

In the early days of near-field optics, small holes were etched in metal layers. This method was not very successful because of the low light intensity. Another setup prevailed, where the original concept, however, cannot be easily recognized. A glass fiber is used whose tip has been formed by pulling or etching. The diameter of the tip can be reduced to 20 nm. An image of such a tip is shown in Fig. 7.45 [282]. In the second step, the outer skin of the glass fiber is metallized, with the exception of the tip. If we strictly followed the concept of Fig. 7.44, two opposite tips would have to be mounted on the front and back of the sample. It is, however, sufficient to operate with one tip since it can be used simultaneously as both input and output. In another setup, the sample itself supplies the local point source so that only one tip is needed for detection.

The setup can be used in different ways. The application as a microscope is obvious. When the probe is moved over the sample, it registers a characteristic optical property as a function of space. The sample can be excited with light, current or electrons while measuring the local luminescence, for example. Since this procedure can be combined with measurements on the basis of the atomic force microscopy, it is very attractive [283]. Vice versa, by fixing the probe to a group of molecules, the local emission spectrum can be measured. Finally, individual molecules have also been successfully detected, cf. Fig. 7.46 [284].

In the meantime, biochemical sensors are a broad field of application. First of all, the tip is dipped into 3-(trimethoxysilyl)propylmethacrylate and afterwards into a fluorescence-containing polymerizing solution. The photopolymerization is caused by an argon laser located at the other end of the glass fiber. This procedure permits the integration of pH-sensitive dye molecules or other biochemical sensor molecules, the fluorescence intensity being a measure for the pH value. Sensors for calcium, sodium, chloride, oxygen, and glucose were manufactured [282].

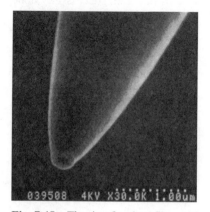

039508   4KV X30.0K 1.00um

Fig. 7.45   The tip of a glass fiber with an aperture of about 100 nm

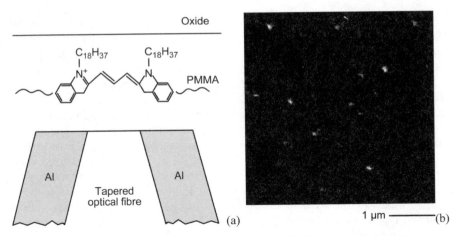

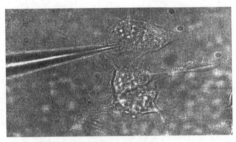

Oxide

C$_{18}$H$_{37}$ C$_{18}$H$_{37}$

PMMA

Al   Al

Tapered
optical fibre

(a)   1 µm ———————(b)

Fig. 7.46   Measurement of a single 1,1'-dioctadecyl-3,3,3',3'-tetramethylindocarbocyanin molecule with a near-field probe. (a) Measurement setup, (b) fluorescence image

7.7.2   Evaluation and Future Prospects

It may be somewhat questionable to group near-field optics into the structuring procedures. There are, however, three reasons: (i) Near-field optics is a means of detecting and resolving natural inhomogeneities or artificially created, nanoscale structures. (ii) To a limited extent, molecules can be photosynthetically created on the glass fiber tip, i.e., in a working area of some 10 nm. (iii) It can be expected that photosynthetic reactions can be induced within suitable molecules in the future. The glass fiber tip serves as a burner to start the reaction.

The production of small and sufficiently stable tip surfaces still encounters difficulties. Materials other than glass or other processing techniques can possibly be used here. Near-field optics, however, offers an enormous potential. From the scientific point of view, it is tempting to measure biological systems such as an individual blood cell with a resolution of 100 nm. This has been successfully done *in vitro* with a fertilized egg cell of a rat [282].

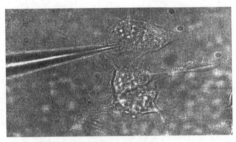

Fig. 7.47   Near-field probe in a vascular smooth muscle cell

As another example, the local calcium content in a vascular smooth muscle cell can be measured [282]. Destruction of the plasma membrane or the cell organelles after the penetration of the near-field probe is not observed. The local fluorescence signal can then be measured (Fig. 7.47).

Also, the excitation and decay times of the signal (e.g., fluorescence) after the application of an external stimulus can be measured (e.g., after rinsing the above-mentioned fertilized cell with diamide). Moreover, it is conceivable that a manipulator of the size of the probe is brought into the cell which is afterwards excited locally. These experiments obviously give us enormous hope for the diagnosis and therapy of diseases.

8   Extension of Conventional Devices by Nanotechniques

8.1   MOS Transistors

The ever progressing and seemingly unstoppable miniaturization of MOS transistors is the essential factor responsible for the progress of nanotechnology. Today, MOS transistors with channel lengths of around 100 nm have already been introduced in the production of memory modules and microprocessors. Further development indicates that MOS technology will be used for silicon transistors with channel lengths down to 25 nm. According to the ITRS, the reduction of the structure sizes is expected to advance as presented in Table 8.1.

It is doubtful whether these dimensions will eventually be achieved. On the one hand, basic production equipment for structure widths below 70 nm line width is still missing. Furthermore, statistical effects suggesting a reduction of the yield have been neglected up to now. On the other hand, any prediction in microelectronics has been exceeded in best time.

8.1.1   Structure and Technology

MOS transistors with dimensions below 70 nm have been introduced in several publications [285, 286]. The conventional technology for the production of these elements is only adapted and optimized there. Basically, no new procedures are used.

Table 8.1   Field data and minimum structure size for MOS components according to the ITRS [1]

Year	1999	2000	2001	2002	2003	2004	2005	2008	2011	2014
Technology, nm	180			130			100	70	50	35
DRAM, nm	180	165	150	130	120	110	100	70	50	35
MPU-Gate, nm	140	120	100	90	80	70	65	45	32	22
Lithography	KrF			KrF-RET ArF			ArF-RET F_2	F_2-RET	EUV IPL EPL	EUV IPL EPL

The following essential process steps for the transition from the micrometer scale to the nanometer regime are changed:

- Adjustment of the gate oxide thickness to a few nanometers
- Reduction of the doping depths to a few nanometers
- Optimization of the spacer width and of the LDD doping (lightly doped drain)
- Optimization of the channel doping
- Introduction of special implantations (pocket implantation)

Gate Oxide Thickness

Due to the tunneling effect, the scaling of the oxide thickness is limited. Below 3 nm oxide thickness, a current flow occurs through the oxide already for small potential differences, which leads, for instance, to unwanted leakage currents in memory modules. Additionally, it causes a rise of the energy dissipation in complex circuits. Transistors with thinner gate oxides can be manufactured, but tunneling currents within the nanoampere range must then be tolerated [287].

The application of thicker gate dielectrics with higher dielectric constants is an alternative. Titanium dioxide and tantalum pentoxide as well as ferro-electrical materials such as barium titanate with oxide-equivalent layer thicknesses below 1 nm are suggested [1]. However, the deposition of homogeneous nonporous layers of these materials is not yet achieved with reproducible results.

Doping Depths

For MOS transistors with channel lengths of a few nanometers, a reduction of the usual doping depths of about 50–100 nm is necessary. A maximum depth of 10 % of the channel length is useful so that doping depths of 3–5 nm will be required in the future. These cannot be obtained in every case with today's common implantation systems.

In silicon technology, donor elements like phosphorus, arsenic, and antimony are available. Even with a small particle energy, phosphorus manifests a relatively high penetration depth into the crystal. Additionally, this element diffuses some nanometers during the activation annealing. Arsenic manifests a high solubility, but as a heavy element, it penetrates only the near surface of the crystal. Diffusion during activation is negligible.

Being the heaviest element, antimony penetrates only a few atomic distances into silicon. However, its solubility is limited. Diffusion is more strongly pronounced than in the case of arsenic. The small lateral straggling of the element under the mask edge due to the small penetration depth is advantageous. Thus, antimony is suitable for the doping of the LDD area between the channel and the highly arsenic-doped drain and source contacts of the n-channel MOS transistor.

Acceptors in silicon are boron, aluminum, gallium, and indium. In order to likewise obtain a doping near the surface in the p-channel transistor, an element as heavy as possible must be selected.

Because of its relatively high activation energy, indium is ruled out. At room temperature, only 10 % of the introduced dopants are electrically active. In connection with the small solubility of the material in the silicon, no suitable doping can be achieved. Gallium exhibits a very high diffusion coefficient both in the oxide and in the silicon. At the typical process temperatures from 750–1050 °C the implanted impurity concentration profiles substantially smear out. Flat dopings are not attainable even with near surface implantation.

The small mass of aluminum already excludes a very near surface implantation. The element exhibits a sufficient solubility in silicon. Moreover, diffusion in silicon is not too high. However, the problem is the mass of the ion in connection with the source material. Due to the interference of the mass number of aluminum with molecular nitrogen and carbon monoxide, no pure ion beam can be produced with today's implantation systems. The extremely corrosive $AlCl_3$ and trimethylaluminum are so far available as source materials. However, both materials do not supply a sufficiently constant and high ion current.

Therefore, only the element boron is left for the doping, which can be implanted as BF_2 molecule. Thus, the mass number important for the doping depth increases to 49 so that with small implantation energy a near surface doping develops. Nevertheless, due to the strong diffusion of boron, arbitrary flat diffusions cannot be produced.

An alternative to the production of flat p-n junctions in the semiconductor is diffusion from doped oxides. Thus, LDD dopings are produced via diffusion out of phosphorus doped spacers annealed for a short time (rapid thermal annealing, RTA) above 1000 °C [288]. This can also be done for p-conducting diffusions for the production of flat boron profiles.

Optimization of the Spacer Width and LDD Doping

In order to obtain as small a field strength as possible in the transistor, an optimization of the spacer width in the range of a few nanometers in connection with the level of the LDD doping is necessary. A high doping underneath the spacers is aimed at in order to obtain a high transistor conductance.

With a high LDD doping, the space charge zone (SCZ) at the p-n junction drain channel expands far into the channel area. A strong drain voltage dependency of the effective electrical channel length (channel length modulation) results. In this case, the spacer width can be selected very small since only a small expansion of the SCZ occurs into the LDD area.

A weak LDD doping shifts the expansion of the SCZ from the channel area into the LDD. The channel length modulation remains negligibly even for small effective electrical channel length.

Transistors of channel lengths below 50 nm frequently use a double spacer technique in order to ensure a more favorable doping profile from the channel to the drain. At the points of contact to the channel, the dopant concentration of the LDD is selected just as high as in the channel but of the opposite conduction type. A stronger doping follows so that the gradient of the spatial dopant concentration remains small. Therefore, only moderate short channel effects occur in these transistors.

As an alternative to the double spacer technique, the electrical characteristics of the transistors can be improved by lateral implantation under the gate electrode. By an implantation of the wafer surface at a large irradiating angle, a "pocket implantation" is laterally applied to the gate electrode which, comparable to the LDD doping, leads to the decrease of the short channel effects.

Correspondingly, the channel area of the transistor requires a doping adjustment. For the suppression of the penetration of the SCZ, a dopant increase between drain and source under the conducting channel is necessary. This can be done by an ion implantation.

8.1.2   Electrical Characteristics of Sub-100 nm MOS Transistors

For channel lengths below 100 nm, parasitic short channel effects are increasingly dominant and are difficult to reduce with the usual countermeasures. Thus, measures such as a further reduction of the gate oxide thickness or the decrease of all doping depths are both technologically as well as physically limited as already described above. While the electrical characteristics of the MOS transistors such as slope and switching speed have been improved in the past by the progressive reduction of the transistor dimensions, a rather opposite trend is to be expected for the sub-100 nm transistors.

Figures 8.1–8.3 are exemplary representations of the input characteristics of three sub-100 nm transistors. While the transistor with a channel length of 70 nm (Fig. 8.1) shows a maximum slope normalized to the channel width W of $g_{m,\mathrm{max}}/W = 60\ \mu\mathrm{S}/\mu\mathrm{m}$ for $V_{DS} = 0.1$ V, this value is reduced to $g_{m,\mathrm{max}}/W = 45\ \mu\mathrm{S}/\mu\mathrm{m}$ for the 50 nm transistor (Fig. 8.2), and even to $g_{m,\mathrm{max}}/W = 24\ \mu\mathrm{S}/\mu\mathrm{m}$ for the 30 nm transistor (Fig. 8.3). This is particularly explained by the doping adapted for these transistors. A very short channel length requires very high channel doping in order to minimize parasitic short channel effects and to counteract the reduction of the threshold voltage by the threshold voltage roll-off. However, the increase of the channel doping leads to a decrease of the carrier mobility, which explains the reduction of the slope with decreasing channel length (and with the same time necessary higher channel doping). On the other hand, increasing flat drain/source doping is necessary which leads to increasing parasitic series resistances and thus additionally to decreasing slopes.

Figures 8.4–8.6 show examples of measured output characteristic fields of sub-100 nm MOS transistors. The represented voltage ranges are adapted to the respective maximum voltage stability (which significantly reduces with decreasing channel length) and thus deviate from each other. Qualitatively, it is shown that the gradient of the characteristics in the saturation regime of the transistor increases with decreasing channel length. According to the ideal transistor equations, the gradient should be zero in this case. Thus, the transistor exhibits an infinite output resistance in the saturation regime, which would be equivalent to an output conductance $g_{DS} = 0$. Due to the parasitic effect of the channel length modulation, however, the output conductance g_{DS} rises with decreasing channel length, which is apparent in a clear increase of the characteristics in the saturation regime. Consequently, the maximum attainable voltage amplification $v_i = g_m / g_{DS}$ of the transistor is reduced.

Dynamic investigations show a trend that the switching speed of sub-100 nm MOS transistors does not increase by the amount that is generally expected. The reasons are the increasing doping gradients which lead to increasing parasitic capacitances. Analyses by a large number of independent scientists show, however, that in the future the delay time in the signal lines of the microchip will dominate and hence the switching times of the transistors do not need to be given much attention any more, contrary to today's conditions [289].

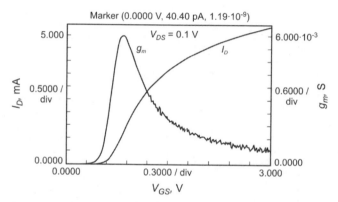

Fig. 8.1 Measured input characteristics of an NMOS transistor with $L = 70$ nm, $W = 100$ μm, and $t_{ox} = 4.5$ nm

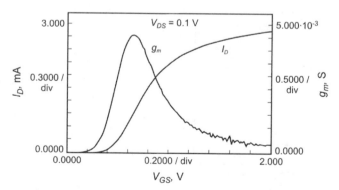

Fig. 8.2 Measured input characteristics of an NMOS transistor with $L = 50$ nm, $W = 100$ μm, and $t_{ox} = 4.5$ nm

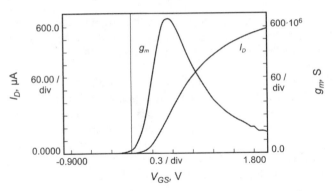

Fig. 8.3 Measured input characteristics of an NMOS transistors with $L = 30$ nm, $W = 25$ μm, and $t_{ox} = 4.5$ nm

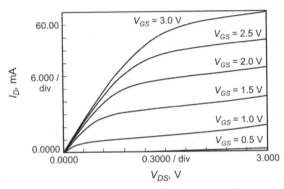

Fig. 8.4 Measured output characteristic field of the NMOS transistor of Fig. 8.1 with $L = 70$ nm

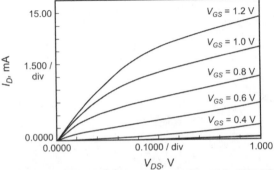

Fig. 8.5 Measured output characteristics of the NMOS transistor of Fig. 8.2 with $L = 70$ nm

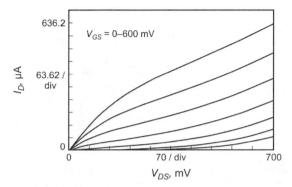

Fig. 8.6 Measured output characteristics of the NMOS transistor of Fig. 8.3 with $L = 30$ nm

8.1.3 Limitations of the Minimum Applicable Channel Length

While neither the static nor the dynamic characteristics of sub-100 nm MOS transistors for channel lengths down to 30 nm prevent a practical applicability in digital circuits, statistically, physically caused fluctuations could become a problem. While statistical fluctuation of the electrical device parameters have been of importance so far mainly for analog circuits, since such deviations limit the accuracy of digital-analog converters and lead to the so-called "offset" via amplification, in the future these statistical fluctuations could also lead to a failure in digital circuits [290, 291]. So far the problems have been largely underestimated and therefore suitable countermeasures are hardly developed and examined.

Two types of statistical fluctuations must be distinguished. There are fluctuations caused by the production process, for example fluctuations of layer thicknesses or of geometrical dimensions. By progress in the processing and by the application of large financial resources for the development and supply of ever more complex manufacturing equipment, these fluctuations have been further lowered in the past. It is very probable that this trend will also continue in the future. But even then, if the tolerances caused by production are completely avoided, fluctuations of the electrical parameters of the transistors can still be observed. The fluctuations are physically caused and therefore cannot be avoided via improved manufacturing equipment.

Figure 8.7 shows experimentally determined distributions of the threshold voltages of MOS transistors with three different channel geometries. Due to a special production process, the transistors show only extremely small scatterings of all geometrical dimensions [292], with which the purely physically caused threshold voltage fluctuations can be observed very well and separately. It is clearly seen that the scattering of the threshold voltage increases significantly with decreasing channel surface.

This physically caused threshold voltage scattering is mainly due to the channel doping of the transistor. It is introduced by ion implantation and thus subject to a

Relative abundance

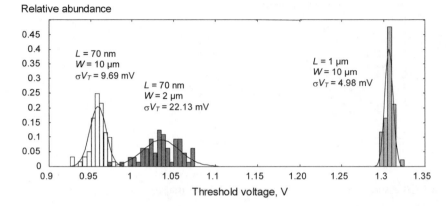

Fig. 8.7 Experimentally determined distribution of the threshold voltage of MOS transistors with three different channel dimensions. The normal distributions calculated from the measured values are represented (solid lines) together with the measured distributions (histograms).

Poisson distribution. According to [293], the statistical threshold voltage standard deviation can be calculated as follows:

$$\sigma V_{Th} = \frac{A_{VTh}}{\sqrt{W\,L}}\,,\ \text{and}\ A_{VTh} = \frac{t_{ox}}{\varepsilon_{ox}}\sqrt{q\,\frac{Q_B}{4}+q^2\,D_i} \qquad (8.1)$$

where Q_B is the charge of the junction depletion region per surface unit: $Q_B = N_A\,W_d$, W_d the depth of the depletion zone, D_i the implantation dose of the threshold voltage, t_{ox} the gate oxide thickness and ε_{ox} the permeability of the gate oxide. Strictly speaking, the A_{VTh} relationship applies only to a homogeneously doped substrate with the doping N_A and to a Dirac-shaped surface doping D_i. The scattering of the threshold voltage obviously increases with decreasing channel surface. This effect can be attenuated by a reduction of the gate oxide thickness t_{ox} within certain limits. The channel doping and its profile have some influence on the scattering of the threshold voltage via D_i and Q_B which is not obvious. An estimate is given in [294]:

$$\sigma V_{Th} \propto \frac{t_{ox}\,\sqrt[4]{N_A}}{\sqrt{W\,L}}\,. \qquad (8.2)$$

It is evident that after minimizing the geometry of the switching element the increase of the channel doping N_A necessary for the reduction of the short channel effects leads to an increase of the scattering of the threshold voltage. Thus, the scattering of the threshold voltage increases with the reduction of the structure due to a gate oxide thickness of limited scalability. This exactly contradicts the demands of ITRS.

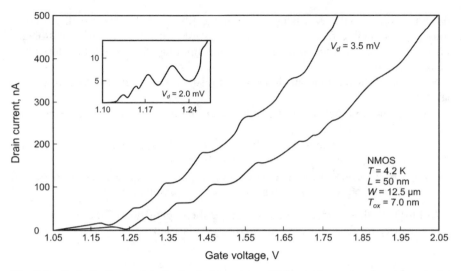

Fig. 8.8 Periodic fluctuations in the drain current with increasing gate voltage

These relations deduced for transistors with channel dimensions above 1 µm could also be verified by measurements [292] and Monte Carlo simulations [295, 296] for sub-100 nm MOS transistors.

The strong increase of the threshold voltage standard deviation could finally limit the minimum applicable channel length. The threshold voltage scattering will drastically increase without countermeasures in the future, but the absolute value of the average threshold voltage must be simultaneously reduced since this one must be adapted to the decreasing operating voltage. By the manufacturing for instance, of a 256 gigabit memory chips, it is then more than doubtful whether all transistors are normally off. It is rather very probable that one or more of the 256 billion transistors on the chip will deviate so strongly in the threshold voltage (here the 6σ or even 7σ deviation must be considered) that they become normally on and thus lead to the failure of the circuit.

Countermeasures could be new circuit concepts, which indicate a certain redundancy, so that the circuit still works in case of the failure of individual transistors. Technologically, a remedy can be created by reducing the channel doping and by adjustment of the threshold voltage via an adapted work function of the gate electrode (work function engineering) [297]. For example, mid band gap materials such as W and Ti or silicon germanium alloys with work functions adjustable via the mixing ratio are suitable [298].

8.1.4   Low-Temperature Behavior

Due to the small channel length of the MOS transistors, quantum effects in these circuit elements cannot be excluded. Measurements of MOS structures with dimensions of 30 nm in the temperature range below 40 K show periodic changes in the drain current with increasing gate voltage [299]. The interval of the oscillations is reproducible from transistor to transistor. An influence of transistor geometry is only ascertained in the height of the amplitude but not in the intervals of the oscillations.

So far, the following models have been consulted for the explanation of this behavior:

- Coulomb blockade: a dependency on the magnetic field is expected, but this does not occur.
- resonance tunneling model: a temperature dependence of the periodic distance should occur, but it is not observed.
- surface states at the gate oxide/silicon junction: these can lead to current fluctuations but never cause the observed periodicity.

In [301], further models are consulted for the explanation of the phenomenon, but the causes for these fluctuations are not yet clarified.

8.1.5   Evaluation and Future Prospects

The MOS technology will presumably be continued up to the year 2012 by the well-known scaling of structure geometry and thus deeply penetrate into the nanometer range. Severe effects which impair device function do not appear for transistor channel lengths down to 25 nm but a reduction of the yield is to be expected due to the statistical distribution of the dopant. Thus, an economical scaling

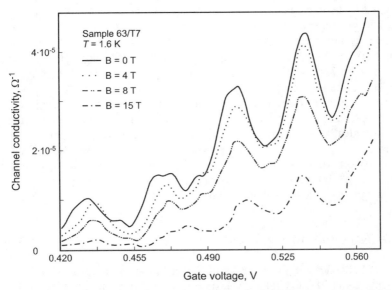

Fig. 8.9   Periodic fluctuations of the conductance for different magnetic field strengths [300]

boundary within the range between 70 and 50 nm channel length results for the MOS technology.

The quantum effects in these devices observed so far are relevant only for very low-temperature operation. Above approximately 50 K, no disturbances in the transistor characteristics are published. It is unknown whether further quantum effects occur below 30 nm channel lengths.

8.2   Bipolar Transistors

8.2.1   Structure and Technology

The bipolar technology uses structures with nanometer dimensions only during the self-adjusting bipolar process. Due to the self-adjustment of the dopings relative to each other, this integrated-circuit technique enables transit frequencies in the range above 40 GHz for pure silicon transistors and up to about 120 GHz for silicon germanium switching elements. For the production, extremely thin epitaxial films of different doping levels are used as collector (100 nm) and base layers (<50 nm) instead of implantation or diffusion. Only the emitter is diffused from a polysilicon layer into the crystal. Both the base and the emitter contacts are self-adjusted relative to each other with the help of spacer structures manufactured with a width of about 50 nm. The design of such a transistor is presented in Fig. 8.10.

This self-adjusting bipolar process is characterized by high critical frequencies (>40 GHz) of the circuit elements in connection with a relatively high packing density. The typical area of the emitter amounts to about $0.15 \cdot 1.5 \ \mu m^2$.

Further increases of the critical frequency are possible with a base from a heteroepitaxially grown crystalline silicon-germanium epitaxial layer which is deposited on a silicon substrate with the molecular beam epitaxy or via MOCVD procedure. With a germanium content around 20 % of the atomic composition, the mobility of the charge carriers rises on the one hand. On the other hand, the germanium doping causes a modification of the band structure and enables an ex-

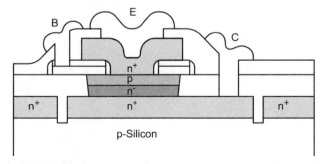

Fig. 8.10   Cross section of a bipolar transistor, manufactured in self-adjusting structural form

tremely narrow and very highly doped base [302]. Correspondingly, manufactured bipolar transistors reach critical frequencies of over 100 GHz.

8.2.2   Evaluation and Future Prospects

Since many typical applications of the bipolar transistors in the high frequency regime are taken over today by MOS transistors, the fields of application of these elements in the future are exclusively within the very high frequency regime. The heterojunction bipolar transistors from SiGe are particularly suitable for this purpose. The nanostructuring of bipolar transistors will lead to a further increase of the critical frequencies, but no substantial technological innovation is to be expected in this area.

9 Innovative Electronic Devices Based on Nanostructures

9.1 General Properties

A simple and generally accepted definition of the term *nanoelectronic device* does not exist. In most cases, however, this term is applied to devices which have an important component that lies in the nanometer scale. When taking the development of the MOS technology as an example, the ambiguity of this definition becomes apparent. Almost since the beginning of MOS technology, the thickness of the gate isolator has been in the nanometer scale (about 100 nm in the 1980s, and less than 10 nm nowadays [1]). MOS transistors had not been considered as nanoelectronic devices until the channel length had reached a value less than 100 nm. This has happened only recently. In the case of the quantum dot laser, all the dimensions of the device exceed the nanometer scale while the quantum dots embedded into the active layer of the laser diode have nanoscale dimensions. Since these quantum dots—the site where radiative recombination takes place—are the most important component of the quantum dot laser, this laser is referred to as a nanoelectronic device.

Applying the definition given above, all the other quantum effect devices can also be considered as nanoelectronic devices, in principle. As an example of this class of devices, we will review the state-of-the-art of the resonant tunneling diode (RTD). It is the most simple quantum effect device with regard to its structure and enables high frequency and ultrafast digital electronic applications. Furthermore, the quantum cascade laser (QCL) is reviewed, which is used as an optoelectronic light emitter with emission wavelengths ranging from the near infrared up to wavelengths as high as 200 μm, thus corresponding to terahertz frequencies.

An important part of this chapter is devoted to the comparison of the properties of nanoelectronic devices and conventional electronic devices.

9.2 Resonant Tunneling Diode

9.2.1 Operating Principle and Technology

With regard to their structure, resonant tunneling diodes (RTD) are probably the simplest devices based on quantum effects in semiconductor nanostructures. The basic RTD device incorporates a double barrier quantum well (DBQW) structure

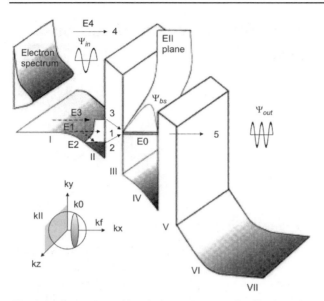

Fig. 9.1 Structure and local electron energy distribution of a resonant tunneling diode with applied bias voltage (from [303])

which is schematically presented in Fig. 9.1 [303]. The contact layers in the areas I, II, VI, and VII consist of a heavily doped semiconductor with a relatively small band gap, for instance, GaAs. The layers III and IV are the tunneling barriers implemented with semiconductors of a relatively large gap and in particular with a large conduction band offset relative to the neighboring regions like AlGaAs. The quantum well layer confined by the two tunneling barriers again consists of a material with a relatively small band gap.

The operating principle can be explained with the help of Fig. 9.1: a local distribution of the electron energy is shown when a bias voltage is applied to a DBQW structure. The energy distribution of the electrons in the heavily doped region I must be described by the Fermi-Dirac distribution, and the electrons in this region are assumed to be in thermal equilibrium. At the boundary surfaces, there are multiple reflections of the electrons due to their wave nature, leading to destructive and constructive interferences as a function of the electron energy. Thus, the tunneling of those electrons is favored which hit the left barrier with an energy E_1 corresponding to the energy E_0 in the quantum well. The tunneling probability decreases drastically with both higher and lower electron energies. However, since the maximum of the electron energy distribution in the regions I and II can be tuned by changing the applied bias voltage, a local maximum (peak) is found in the current-voltage characteristics of the resonant tunneling diode, followed by a local minimum (valley). In Fig. 9.2, this is shown for an InGaAs/AlAs based RTD at 300 and 77 K [304].

Even at room temperature, a region with a negative differential resistance (NDR) can be clearly identified with a peak-to-valley ratio better than 20. Such

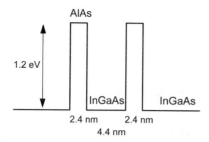

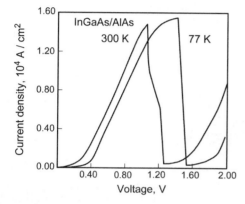

Fig. 9.2 Conduction band diagram and current-voltage characteristics at 77 and 300 K of an AlAs/InGaAs RTD (from [304])

characteristics suggest both bistable and astable applications, and, indeed, the most common applications of RTDs are microwave oscillators operating at extremely high frequencies and very fast digital electronic circuits.

It should be mentioned that a negative differential resistance is also found in InAs/AlSb/GaSb resonant tunnel structures with AlSb barriers and GaSb quantum wells. In this case, however, the special band structure favors electron tunneling between the energy levels within the valence band of the AlSb barriers and the energy levels within the conduction band of the GaSb quantum well layer. Consequently, the resulting device is referred to by the name *resonant interband tunnel diode* (RITD) [305, 306].

As in the area of optoelectronics, it is attempted to replace III-V materials with silicon for the production of resonant tunneling devices because of lower costs and the possibility of integrating them with VLSI silicon ICs. As an example, the schematic structure and band diagram of an RTD with silicon quantum wells and CaF_2 barriers are depicted in Fig. 9.3. Alternatively, Si/Ge [308, 309] or silicon-on-insulator (SOI) technology [311] can be employed. The layer sequence and RTD structure of an SOI technology-based RTD are shown in Fig. 9.4. The well material of 2.5 nm thickness is confined by two ultrathin (2 nm) buried SiO_2 layers. For this device, a negative differential resistance is observed at operating temperatures of up to 100 K.

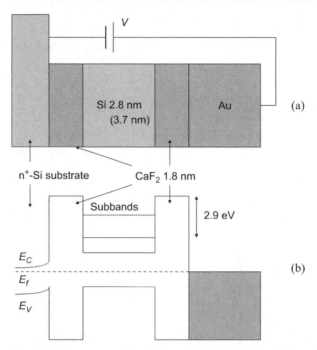

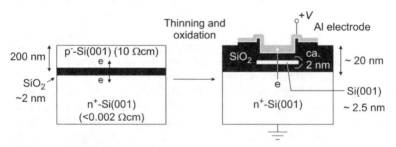

Fig. 9.3   Structure and band diagram of a Si/CaF$_2$ double barrier RTD (from [307])

Fig. 9.4   Layer sequence and device structure of a Si/SiO$_2$ double barrier RTD (from [310])

9.2.2   Applications in High Frequency and Digital Electronic Circuits and Comparison with Competitive Devices

Due to the rapid progress of microwave transistors regarding their high frequency behavior, 2-terminal devices were ousted from high frequency oscillator applications for frequencies below 30 GHz. This progress led to transistor cut-off frequencies of 350 GHz for InP/InGaAs heterobipolar transistors (HBT) [311], 42 GHz for SiGe FETs [312] and 85 GHz for GaN/AlGaN high electron mobility transistors (HEMTs) [313]. For higher frequencies (30 GHz–1 THz), IMPATTs,

Gunn diodes, and resonant tunneling diodes are still being considered for applications as microwave oscillators [314].

For transit time diodes (IMPATTs), a maximum oscillator frequency of 400 GHz has been obtained with an output power of 0.2 mW [314], and at 44 GHz an output power of 1 W has been measured [315]. Similar values can be obtained by means of transferred electron devices (TEDs), also known as Gunn diodes. In the latter case, an output power of 34 mW at 193 GHz [316] and of about 96 mW at 94 GHz [317] has been achieved. Despite their somewhat lower performances at frequencies below 100 GHz, Gunn diodes are an important alternative to IMPATT diodes due to their low noise operation.

In comparison to the two devices treated so far, lower output powers are achieved with resonant tunneling diodes. At 30 GHz and 200 GHz, for example, output powers of about 200 mW and 50 µW, respectively, have been reported [314]. However, with InAs/AlSb RTDs, a record frequency of 712 GHz has been obtained, achieved by an InAs/AlSb RTD with an output power of 0.3 µW [318]. The limitation of the maximum output power of RTD based oscillators is mainly caused by the relatively high series inductance [314]. A further advantage of RTDs compared to IMPATT and Gunn diodes is the fact that they can be easily integrated with other electronic devices, like modulation doped field effect transistors (MODFETs) and heterobipolar transistors [309]. Thus, they are attractive for microwave integrated circuits (MMICs) even at moderate frequencies in the GHz range. Another reason of making resonant tunneling diodes appealing for digital circuit applications is the possibility of implementing very compact logical circuits since the number of active devices can be reduced as compared to conventional digital electronic circuits. In Table 9.1, the number of active devices required for the implementation of several digital functions using RTDs in comparison with TTL, CMOS, and ECL technology is listed [319].

Resonant tunneling diodes intrinsically have very short switching times of about 1.5 ps. As stated earlier, they can be easily integrated with ultrafast transistors like MODFETs and HBTs [320]. Moreover, they can operate at room temperature, which is a clear advantage as compared to a superconducting integrated circuit—another possible competitor for the implementation of ultrafast digital electronic circuits [321]. A further advantage is the possibility of quite easily implementing multi-value-logic systems using multi-peak resonance RTDs. In Fig. 9.5, the circuit of a digital counter implemented with just three HBTs and one RTD and the corresponding output characteristics of the transistor Q1 with the multi-peak RTD as a load are depicted [319].

Table 9.1 Number of active devices required for the implementation of digital functions using RTD, TTL, CMOS, and ECL technologies (from [319])

Logical function	TTL	CMOS	ECL	RTD
bistable XOR	33	16	11	4
9-state memory	24	24	24	5
NOR2 + flip-flop	14	12	33	4
NAND2 + flip-flop	14	12	33	4

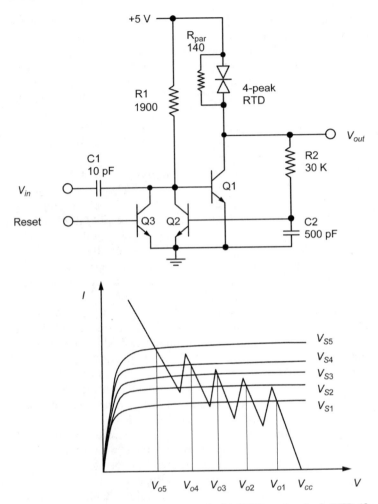

Fig. 9.5   Counting circuit implemented with HBTs and a single RTD (from [319])

Other examples of digital circuits are a 50 GHz frequency divider fabricated with one resonant tunneling diode and one HEMT [322] as well as NAND and NOR gates consisting of a single resonant tunneling bipolar transistor with an integrated RTD structure (RTBT) [323]. Besides digital circuit applications, analog applications have also been proposed, e.g., analog/digital converters using an RTD as a multi value comparator [324]. In optoelectronic applications, the intrinsic bistability of RTDs is used in particular. The combination of an RTD grown on top of a multi-quantum-well electro-optic modulator exhibited bistable operation at room temperature with switching powers in the mW range [325]. The monolithic integration of an InAlAs/InGaAs RTD with an InGaAs/InGaAsP traveling wave photodiode on the same InP substrate enabled the production of an optoelectronic flip-flop operating at a clock rate of 80 Gb/s [326].

Future nanoelectronic digital circuits could be manufactured using chemical self-organized growth of quantum dot arrays. In this case, it would be much easier to use two-terminal devices rather than transistor-like three-terminal devices. The RTD is an attractive candidate for such nanodevice circuits because it combines the ability to implement complex logic functions with a relatively simple structure [327]. A first step in this direction is the demonstration of NDR behavior at 4 K due to resonant tunneling through single InAs quantum dots, obtained by self-organized growth [328].

9.3   Quantum Cascade Laser

9.3.1   Operating Principle and Structure

The quantum cascade laser (QCL) is the equivalent to the quantum well infrared photo detector (QWIP) with regard to the optical emitter. The emission of light quanta, however, is not based on *inter*band transitions as in the case of the classical semiconductor laser but on *intra*band transitions. More precisely, radiating transitions between different energy levels within individual neighboring quantum wells are used in the case of the QCL. Therefore, it is also possible to construct optoelectronic devices operating in the far infrared (3.8–200 μm) with material systems based on semiconductors with a relatively large band gap. The advantage in this case is the possibility of using GaAs/GaAlAs and InP/InAlAs and to benefit from their well-developed technology. The energy difference between the energy levels in the quantum well does not only depend on the barrier height of the quantum well but also on its width. Hence, the possibility of changing the emission wavelength without changing the barrier and quantum well material, but only by varying the thickness of the quantum well layers is very interesting. This method of *band gap engineering* enables the implementation of emitters with very different emission wavelengths with just one technology.

Moreover the quantum cascade laser is a unipolar device, which means that the emission is in general only based on electronic transitions in the conduction band. Since the charge carriers do not recombine during radiative transition, they can be used several times for the light emission by repeating the basic structure. This means that quantum efficiencies >1 can be achieved. In state-of-the-art quantum cascade lasers, the basic units are repeated between 20 and 40 times [329].

Despite the operating principle being similar, the quantum cascade laser exhibits a substantially more complicated structure in comparison to the quantum well infrared photodetector (QWIP) [330]. This difference is explained by an additional condition necessary for the operation of the laser: the population inversion between the two energy levels. Thus, the higher energy level must have a higher electron concentration than the lower level. In principle, the operation of the laser can be described with the help of the conduction band profile of a basic unit of this laser (Fig. 9.6) [331]. In the case illustrated, the barriers are made of AlInAs layers, and the quantum wells consist of GaInAs. Electrons tunnel from the injector into the electronic level E_3 and fall on level E_2 by sending out light quanta with

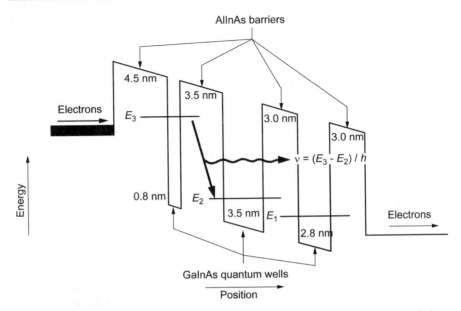

Fig. 9.6 Conduction band profile in the area of the active layer of a quantum cascade laser (from [331])

$v = (E_3 - E_2)/h$, followed by a non-radiative transition from energy level E_2 to level E_1.

To achieve population inversion and laser operation, the electron tunneling rate from the two levels E_2 and E_1 into the neighboring conduction band must be higher than the tunneling rate from the level E_3. Therefore, an electronic Bragg reflector consisting of a semiconductor superlattice is inserted into the QCL structure behind the active layer. This structure is depicted in Fig. 9.7a. The probability of transmission of the Bragg reflector's electrons as a function of the energy position relative to the conduction band minimum can be seen in Fig. 9.7b [332]. As can be clearly seen, a forbidden band (mini-gap) develops and takes effect within the energy region of the level E_3. The tunneling probability from the level E_3 is two orders of magnitude lower than from the levels E_2 and E_1, whose energies correspond to the energy region of the mini-band in the Bragg reflector section of the laser. Thus, the necessary condition for population inversion is ensured.

The first quantum cascade laser ever devised achieved a few mW optical power at cryogenic temperatures under pulsed operation conditions. As an example of these early QCLs, the emission spectrum and the emitted power as a function of the laser current of an InGaAs/InAlAs Fabry-Perot quantum cascade laser with an emission wavelength around 4.5 μm are depicted in Fig. 9.8 [332].

An overview of the current developments concerning the optical power and the covered wavelength range of quantum cascade lasers is given in Table 9.2. At present, an emitted optical power of more than 1 W near room temperature [335]

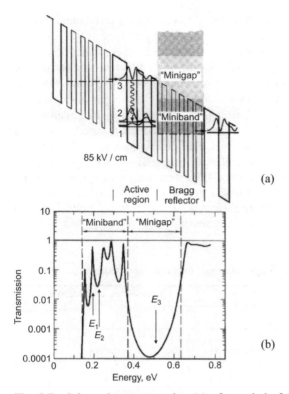

(a)

(b)

Fig. 9.7 Schematic representation (a) of a period of the quantum cascade laser and (b) energy dependent transmission probability for electrons of the Bragg mirror (from [332])

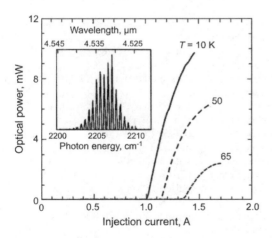

Fig. 9.8 Optical power as a function of laser current and optical emission spectrum of a Fabry-Perot InGaAs/AlInAs quantum cascade laser (from [332])

and of up to 6 W at a temperature of 80 K [333] have been achieved under pulsed operation. It is also interesting to note that a wavelength range from 3.8 up to 24 μm can be covered with QCLs using only three different combinations of active layer materials. For example, this wavelength range is interesting for optical data transmission within the earth's atmosphere, where there are optimal transmission conditions in the two spectral range windows from 3.5 to 4 μm and from 8 to 10 μm. It should be noted that the full infrared spectral range can be covered with III-V compound semiconductor lasers. Conventional semiconductor lasers based on electron-hole recombination cover the wavelength range up to 3.25 μm [345], while QCL lasers cover the longer wavelength range. Besides the Fabry-Perot (FP) type lasers, also distributed feedback (DFB) quantum cascade lasers are developed today. Due to their narrower emission spectrum, these are better suited for spectroscopic and data transmission applications. In Fig. 9.9, the optical power-laser current characteristics and the optical emission spectrum of a DFB InGaAs/

Table 9.2 Optical powers and wavelength ranges of quantum cascade lasers

First author (Year)	Active layer material	Electronic (optical) structure	Max. optical power (at temperature)	Emission wavelength, μm
Yang et al. [333] (2002)	InAs/GaInSb/ AlSb	MQW (Fabry-Perot)	pulsed 6 W (80 K) pulsed 1 W (150 K)	3.8
Yang et al. [334] (2002)	GaInAs/AlInAs	MQW (Fabry-Perot)	pulsed 900 mW (77 K), pulsed 240 mW (300 K)	5.0
Hofstetter et al. [335] (2001)	GaInAs/AlInAs	undotiertes SL (DFB)	pulsed 1,15 W (273 K) pulsed 92 mW (393 K)	5.3
Scamarcio et al. [336] (1997)	GaInAs/AlInAs	MQW (Fabry-Perot)	pulsed 750 mW (80 K) pulsed 200 mW (210 K)	8.0
Faist et al. [337] (2000)	GaInAs/AlInAs	nipi SL (Fabry-Perot)		8.8
Page et al. [338] (1999)	GaAs/AlGaAs	MQW (Fabry-Perot)	pulsed >1 W (77 K)	9.7
Hofstetter et al. [339] (1999)	GaInAs/AlInAs	MQW (DFB)	pulsed 80 mW (300 K)	10.2
Tredicucci et al. [340]	GaInAs/AlInAs	MQW (Fabry-Perot)	CW 75 mW (25 K) CW 8 mW (80 K)	11.1
Anders et al. [341] (2002)	GaAs/AlGaAs	MQW (Fabry-Perot)	pulsed 75 mW (78 K) pulsed 2 mW (300 K)	12.6
Rochat et al. [342] (2001)	GaInAs/AlInAs	*chirped* SL (Fabry-Perot)	pulsed 400 mW (210 K) pulsed 150 mW (300 K)	16.0
Colombelli et al. [343] (2001)	GaInAs/AlInAs	undoped SL (Fabry-Perot)	pulsed 14 mW (70 K)	19.0
Colombelli et al. [344] (2001)	GaInAs/AlInAs	undoped SL (Fabry-Perot)	pulsed 3 mW (70 K)	24.0

AlInAs quantum cascade laser are depicted. At room temperature, this laser emits around 7.8 µm with an optical power of up to 10 mW [346]. Furthermore, the emission wavelength can be tuned between 7.75 and 7.85 µm by varying the temperature from 100 to 300 K, which is of interest to laser spectroscopy.

For quantum cascade lasers with emission wavelengths below 4 µm, InAs/GaInSb/AlSb is preferred as active layer material [333]. With regard to the type of the electronic structure, nipi-superlattices (layer sequence of n-type, intrinsic, p-type, intrinsic) [337] or intrinsic superlattices [335, 343, 344] have been used as the active layer of QCLs, besides the widespread multi-quantum-well (MQW). Thus, lasers with a high optical power in the spectral range around 5 µm [335] and lasers with emission wavelengths up to 24 µm [344] have been developed.

Recently the production of GaAs/GaAlAs quantum cascade lasers with emission at wavelengths between 100 and 200 µm has been reported [347, 348, 349]. These monochromatic Terahertz emitters close the gap between electronic and optical oscillators in this wavelength range. Terahertz emission has interesting applications in biomedical imaging [350]. As shown in Fig. 9.10, a QCL with an emission frequency of 4.44 THz and a power of more than 2 mW has been reported [348]. This is a much higher value of the emitted power as compared to

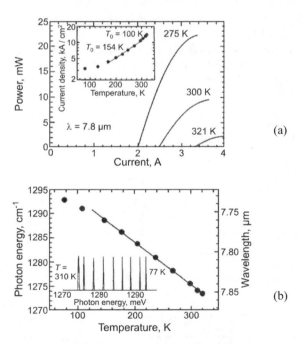

Fig. 9.9 (a) Optical power as a function of laser current. Inset: temperature dependence of the lasing threshold current density. (b) Temperature dependence of the emission wavelength of a distributed feedback InGaAs/AlInAs quantum cascade laser. Inset: optical emission spectra for different temperatures between 77 and 300 K (from [346])

pure electronic Terahertz generation using harmonic generation. It should be mentioned, however, that for broadband Terahertz generation using relativistic electrons in a particle accelerator emission powers greater than 20 W have been reported [351].

First results with regard to electroluminescence in Si/SiGe cascade emitters [352] offer a good chance of manufacturing of silicon based quantum cascade lasers in the future [353].

9.3.2   Quantum Cascade Lasers in Sensing and Ultrafast Free Space Communication Applications

Due to the short electron lifetime, the quantum cascade laser is an optical emitter that can be—in principle—modulated very fast. Thus, theoretical modulating frequencies higher than 100 GHz can be achieved. These frequencies are higher than those of lasers used in optical fiber communication systems at wavelengths of 1300 or 1550 nm. In first data transfer experiments, data rates of up to 5 Gb/s have been achieved using QCLs but still over short distances of maximum 350 m [354, 355, 356].

A further interesting application for lasers with emission in the far infrared is the trace analysis of gases in the atmosphere. For this application, a tunable monomode laser is needed, like the DFB quantum cascade lasers. Methane, nitrogen oxide, ethanol, and the different isotopes of water have been detected in ambient air by means of a sensing system using a quantum cascade laser as emitter and a mercury cadmium telluride (MCT) detector as receiver [357]. In a further step, there is interest in replacing the MCT detector with a quantum well infrared

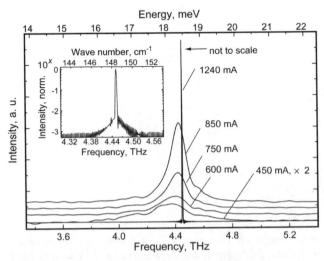

Fig. 9.10   Optical emission spectra of a Terahertz Fabry-Perot GaAs/GaAlAs quantum cascade laser for various laser currents (from [348])

photodetector, using the same material system as for the production of the quantum cascade laser, and possibly to integrate emitter and detector on the same chip. Another alternative would be to use the QCL itself as photodetector. This functionality has already been demonstrated [358].

9.4   Single Electron Transistor

9.4.1   Operating Principle

The single electron transistor (SET) is an example of an electronic device where a final limit of electronics has already been reached: the switching of a current carried by just one electron. The operating principle can be understood with the help of Fig. 9.11 [359].

In principle, the SET consists of two tunnel contacts with associated capacitances (C_s and C_d) and an intermediate island to which an operation voltage (V_g) is capacitively coupled via C_g. By varying V_g, the electrical potential of the island can be changed. An insulator being embedded between two electrical conductors or semiconductors partly loses its insulating characteristics for a layer thickness which is in the lower nanometer range (about 5 nm) due to charge carrier tunneling. Thus, the probability of the transmission of charge carriers increases exponentially with decreasing thickness of the insulator layer. On the one hand, this quantum mechanical effect limits the further miniaturization of the classical MOS transistor due to increasing gate oxide leakage. This leads particularly to stability problems due to the charge carrier capture within the gate oxide and thus to a non-tolerable displacement of the transistor characteristics during prolonged operation.

On the other hand, this tunneling current enables the operation of the single electron transistor. Besides the capacitances C_s and C_d, electrical conductivities $1/R_s$ and $1/R_d$ can also be associated with the two tunnel contacts due to the tunneling current. Under which circumstances will a current flow between drain and source if an external voltage V_d is applied between these contacts? Let us assume that there are already n electrons on the island. One further electron can tunnel through the left barrier, if it has a charge energy of

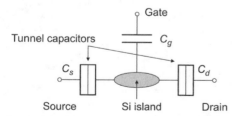

Fig. 9.11   Schematic representation of the double barrier structure of a single electron transistor (from [359])

$$(n+1)\frac{e^2}{2(C_s + C_d + C_g)} \tag{9.1}$$

At the same time, this electron gains energy (ΔE) by its transfer from the left electrode to the island due to the energy difference between these two points:

$$\Delta E = e\frac{V_d\, C_s + V_g\, C_g}{C_s + C_d + C_g} \tag{9.2}$$

Therefore, the electron will only arrive on the island if the total energy difference is negative. In the opposite case, a coulomb blockade is given which can only be overcome by a further increase of the applied voltage V_d. A possible implementation of a SET is shown in Fig. 9.12. The basic element is an ultrathin silicon-on-insulator (SOI) film. Thickness modulation due to anisotropic etching of the center part of the SOI film results in the formation of a series of quantum dots [360]. The gate contact is formed by polysilicon deposition on the upper oxide layer. The resulting current-voltage characteristics of such a device are presented in Fig. 9.13. As can be seen, characteristic levels are formed with constant gate voltage and varying V_d (Fig. 9.13a), while the SET current oscillates at a constant voltage V_d and a varying gate voltage (Fig. 9.13b). This oscillation can be explained by the fact that tunneling from the right contact to the island during increasing gate voltage is also possible and the transistor is again switched into the blocking state. This process is repeated by further increasing the gate voltage and leads to the observed oscillations of the electrical current.

The plots presented in Fig. 9.13 have been taken at room temperature. As a condition for the operation of the SET, the charge energy

$$E_C = \frac{e^2}{2(C_s + C_d + C_g)} \tag{9.3}$$

should be larger than the thermal energy $k\,T$. This means that the maximum operating temperature decreases linearly with increasing device capacitance. While a capacitance of about 1 aF is tolerable at room temperature, this value rises to

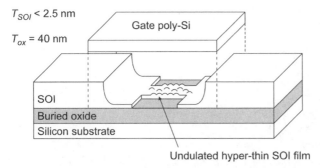

$T_{SOI} < 2.5$ nm

$T_{ox} = 40$ nm

Gate poly-Si

SOI

Buried oxide

Silicon substrate

Undulated hyper-thin SOI film

Fig. 9.12 Schematics of a SOI technology based single electron transistor (from [360])

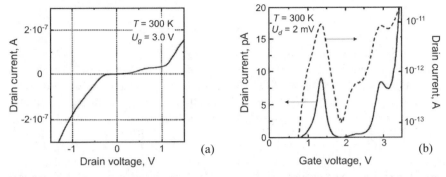

Fig. 9.13 Current of a SET measured at room temperature (a) as a function of the applied drain-source voltage and (b) as a function of the applied gate-source voltage (from [360])

about 60 aF at 4.2 K. Ie., the smaller the device dimensions and thus the electrical capacitances, the higher the maximum allowed operating temperature.

9.4.2 Technology

Listing all technologies used today for the fabrication of single electron transistors would be beyond the scope of this book. Instead some structures which have been implemented so far are specified in Table 9.3 by naming the used island and barrier materials and the respective manufacturing methods. The more interested reader is kindly referred to the indicated quotations where these technologies are described in more detail. At the same time, a characteristic energy E_a is listed in this table. As a rule of thumb, it should be considered that the maximum allowed operating temperature (T) can be estimated from the relation:

Table 9.3 Data on selected single electron transistor technologies

Materials (Island; Barrier)	Fabrication methods	E_a, meV	Reference
Al; AlO_x	Evaporation by means of an e-beam produced mask	23	[363]
CdSe; Organic	Nanocrystal bond to structured gold electrodes	60	[364]
Al; AlO_x	Evaporation on a structured Si_3N_4 membrane	92	[365]
Ti; Si	Metal evaporation on a structured Si substrate	120	[366]
Carboran molecule	E-beam structured thin layer gate + STM electrode	130	[367]
Si; SiO_2	E-beam structuring + oxidation on SIMOX layer	150	[368]
Nb; NbO_x	Anodic oxidation by means of STM	1000	[369]

$$kT < \frac{E_a}{10} \tag{9.4}$$

with k being the Boltzmann constant.

As can be seen, mostly metal/metal oxide and semiconductor/insulator systems are used for the fabrication of SETs. The growing use of organic layers and molecules is also remarkable. Regarding SET manufacturing, evaporation techniques using electron-beam structured masks and structuring with the scanning tunneling microscope (STM) are predominant. Single electron transistors have been fabricated using superconducting Nb/Al structures with AlO_x barriers [369], self-organized growth of GaN quantum dots [370] or Co nanoparticles [371]. The most advanced SETs, however, are devices manufactured by classical microstructuration techniques and based on silicon or III-V semiconductor technology. Coulomb blockade oscillations have been observed in multi-gate SET structures using GaAs/InGaAs/AlGaAs [372] or AlGaAs/GaAs/AlGaAs [373] as semiconductors. Integrated structures of more than one SET have already been produced using silicon nanostructures. As an example, the production of a SET employing the so-called *PADOX* technique is illustrated in Fig. 9.14. This technique is based on the thermal oxidation of silicon quantum wires with a trench structure. It has enabled the fabrication of pairs of single electron transistors with good reproducibility. As a first step toward more complex integrated SET circuits, this technology has also enabled the production of simple inverters, working at a temperature of 27 K [374].

The idea of Coulomb blockade based devices is not limited to single electron transistors, but can be extended to hole transport based devices. This has been

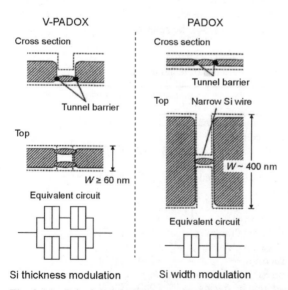

Fig. 9.14 Principle of SET fabrication based on the PADOX and V-PADOX technology (from [374])

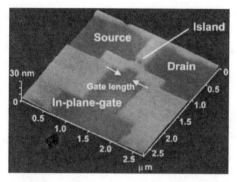

Fig. 9.15   Atomic force microscope image of a diamond single hole transistor
(from [375])

demonstrated by the manufacturing of a single hole transistor (SHT) based on the
modification of a hydrogen terminated diamond surface by means of an atomic
force microscope (AFM). An AFM image of such a transistor is depicted in Fig.
9.15. The bright local oxidized regions can clearly be distinguished from the dark,
hydrogen terminated regions (including the SHT island). This SHT showed the
typical Coulomb oscillations of the drain current with variations of the gate voltage at a temperature of 77 K.

9.4.3   Applications

The main fields of application of the single electron transistor are sensor technology, digital electronic circuits, and mass storage.

As already shown in the preceding section, the SET reacts extremely sensitively to variations of the gate voltage V_g if the voltage V_d is adjusted as Coulomb
blockade voltage, so that an obvious application is a highly sensitive electrometer
[361].

As already mentioned, the Coulomb blockade is only effective if the thermal
energy is lower than the charge energy of the island. Therefore, the differential
electrical conductivity within this area is also a measure for the ambient temperature and enables the use of the SET as a temperature probe, particularly in the
range of very low temperatures [376].

Moreover, the SET is a suitable measurement setup for single electron spectroscopy. For this purpose, the island of the SET structure, for instance, can be
taken as an individual quantum point.

Apart from the applications as sensors, further application as direct current normal is interesting. Since exactly one electron is transported in each period when
applying an alternating voltage of suitable amplitude to the gate of the SET, the
current between source and drain which flows through a single electron transistor
is directly proportional to the frequency of the applied alternating voltage V_g.
Since frequencies can be measured with high accuracy, a precise direct current
measurement standard can thus be implemented [377].

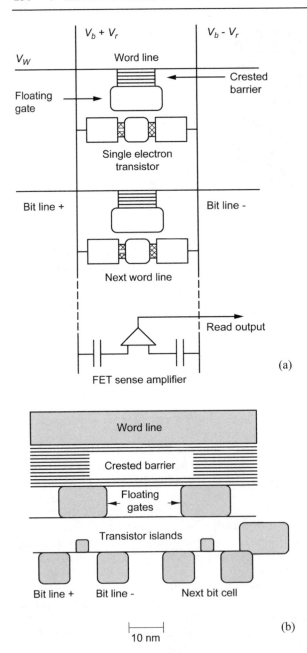

Fig. 9.16   (a) Principle and (b) implementation of a SET-FET hybrid circuit (from [361])

The application of the SET as a switch and memory in digital electronics, operating at room temperature, has found great interest. However, there are some principal problems: as stated earlier, the operation of a single electron transistors at room temperature requires extremely small island capacities of about 1 aF and thus structure widths around 1 nm. Today, this can be achieved in single structures, but the technology necessary for the production of complex digital circuits with these dimensions is not yet mature. Furthermore, there is a tradeoff between the operation of a SET at room temperature and the operation at higher frequencies. Very low device capacitances are obtained with small dimensions, but the electrical resistances associated with the tunneling barriers increase and reach values in the $M\Omega$ range. As a result, the critical frequencies of the SET are limited by relatively large RC constants. Despite all these difficulties, a variety of digital logic functions, including AND and NOR gates, has been obtained with the SOI technology based single electron transistor operating at room temperature (Fig. 9.11).

The potential use of the SET as electronic mass memory has been discussed previously. Here, the number of electrons on a neighboring conducting island containing the stored information could be queried by means of a SET. Memory densities around 10^{12} bit/cm^2 could be reached, which is some orders of magnitude higher than the values achieved by MOS memories today. The main problem consists in disturbing background charges, for example caused by charged impurities in insulating layers, which may induce mirror charges on the island of the SET [378].

Two different concepts to use the SET for data storage should be mentioned: in the first concept—a SET-FET hybrid approach—up to 100 SET based memory cells are read out by a field effect transistor (FET) based amplifier [361]. This type of memory requires—similar to the conventional DRAM operation—a refreshing of the memory content after each reading. It has been estimated that this approach can give memory densities up to 100 Gbit/cm^2 at room temperature. The recording procedure, using a Fowler-Nordheim type tunneling with a typical delay time of approximately 10 ns, is relatively slow. The functionality can therefore be compared to that of an EEPROM. An illustration of the SET-FET hybrid concept is shown in Fig. 9.16 [361]. It is interesting to note that this type of memory is not sensitive to background charges and requires structure lengths of about 3 nm.

In the second concept, it has been suggested that today's dominating magnetic recording technique could be replaced by electrostatic storage (ESTOR) [361]. As presented in Fig. 9.17, the information is kept in loaded grains in a memory layer separated by means of tunnel barriers and a metallic layer from an insulating substrate. The process of writing and reading takes place by using a head floating about 30 nm over the surface with the island of the SET at the top. Memory densities of approximately 1 Tbit/inch^2 are expected. This would be about 30 times higher than the theoretical limit calculated for magnetic memories. Similar memory densities have been estimated for the "Millipede" memory, based on a thermomechanical data storage concept that uses an atomic force microscope cantilever array to read and write a thin polymer film [380].

Similar to the case of the resonant tunneling diode, the SET is a good choice for multi-value logical circuits due to the periodical oscillations in the current-voltage characteristics. As an example for the implementation of such a device, the combination of a SET, fabricated with the already mentioned PADOX process, with a conventional MOSFET resulted in a complex current-voltage characteristics with multiple hysteresis [381]. This is one example that silicon based SETs are very attractive for digital applications because they can easily be interfaced to conventional electronics.

As an example of an analog application of the single electron transistor, the fabrication of a radio frequency mixer has been reported that uses the nonlinear gate voltage-drain current characteristics of the SET for the fabrication of a homodyne receiver operating at frequencies between 10 and 300 MHz [382].

9.5   Carbon Nanotube Devices

9.5.1   Structure and Technology

The nanoelectronic devices presented so far are based on the "classical" materials—silicon and III-V compounds. In this chapter we present recent results on carbon nanotube (CNT) based devices that combine new developments in material science with innovative nanostructuring techniques. As demonstrated in Fig. 9.18, carbon nanotubes are made out of a network with the basic unit being six carbon atoms in ring configuration and arranged in form of cylinders. The electronic structure of the carbon nanotubes as represented by the band diagrams in Fig. 9.18

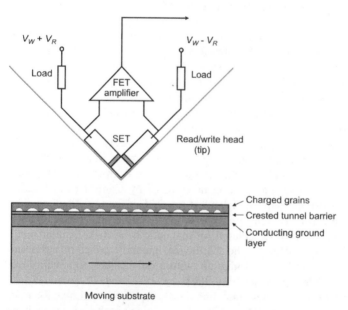

Fig. 9.17   Concept circuit of a writing/reading head for the electrostatic information storage by means of a SET (from [361])

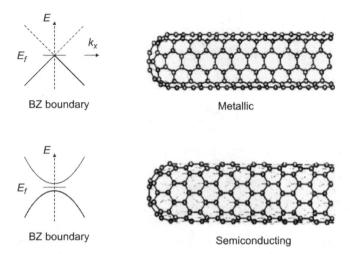

Fig. 9.18   Structure and electronic band diagram of metallic and semiconducting carbon nanotubes (from [383]). Note the different orientation of the rings.

critically depends on the geometry of the interconnection between the carbon rings, resulting either in metallic or in semiconducting behavior [383].

The growth of the cylinders with diameters in the nanoscale range is generally induced by the use of catalytic elements such as iron, molybdenum, and cobalt. The most common growth techniques are arc-discharge [384, 385], laser-assisted deposition [386], and plasma-enhanced chemical vapor deposition (PECVD), using a methane plasma at relatively high temperatures [387]. Depending on the growth parameters, the deposition processes result either in the formation of multi wall nanotubes (MWNTs) or single wall nanotubes (SWNTs) [383].

The particular interest in this new material is due to reports of very low specific resistivities for metallic carbon nanotubes [388] and on high hole mobilities for semiconducting nanotubes [383, 389]. The low density of surface states can physically explain these interesting electronic properties. The material forms a two-dimensional network of carbon atoms without the presence of dangling bonds. When assembling in cylindrical form the problem of the usually enhanced recombination at the edges of the semiconductor can be avoided [383].

First applications of metallic CNTs are wiring of microelectronic circuits and the use as field emitters for high resolution flat panel displays. As an example of the latter application, the manufacturing of a gated 3×3 field emitter cathode array (FEA) [390] is depicted in Fig. 9.19. A silicon surface is covered with a 1-nm thick iron layer as catalyst on top of a 10-nm thick aluminum layer and subsequently with a SiO_2 layer. After deposition of the molybdenum gate electrode and the opening of the single cathode windows by reactive ion etching, metallic multi wall carbon nanotubes are grown on top of the Al/Fe metallization as cathode electrodes using a CVD process with an acetylene plasma at 900 °C. To give an idea of the dimensions of the device: the lengths of the white marks are 50 μm in Fig. 9.19g and 2 μm in Fig. 9.19h.

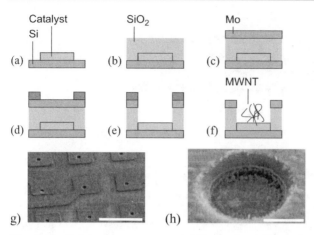

Fig. 9.19   Fabrication process and SEM images of a carbon nanotube based gated field emitter array (from [390])

9.5.2   Carbon Nanotube Transistors

The above-mentioned very high values of the charge carrier mobilities in semiconducting carbon nanotubes together with the small device dimensions make CNT based devices very interesting for microelectronic applications. So far field effect type transistors have mostly been implemented [389, 390, 391] because carbon nanotubes exhibit very high hole mobilities in particular. It should, however, also be mentioned that first experiments to realize a bipolar p-n-p transistor were successful [393].

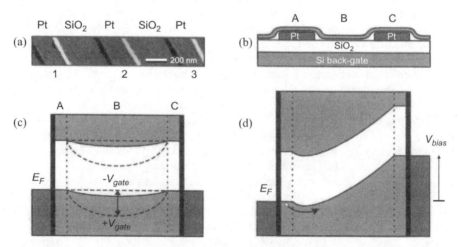

Fig. 9.20   (a) AFM image, (b) schematic structure, and band diagrams without (c) and with (d) applied source-drain voltage of a TUBEFET (from [389])

In Fig. 9.20, the structure and the atomic force microscope (AFM) image of a so-called *TUBEFET* are shown. On a silicon wafer covered with a thermal SiO_2 layer, that serve as backside gate and gate insulator respectively, platinum (Pt) electrodes are deposited that form the source and drain contact. Subsequently the single wall carbon nanotubes are deposited or grown connecting the two platinum contacts. The room temperature characteristics of this TUBEFET are illustrated in Fig. 9.21. For positive voltages applied between drain and source contact, a clear threshold voltage for conduction has been observed whose value increases with increasing gate voltage. For negative voltages applied between drain and source, ohmic behavior has been found. Furthermore it can be seen that a 10 V change of the gate voltage results in a variation of the channel conductance of more than six orders of magnitude.

One of the main problems regarding the fabrication of integrated circuits using CNT transistors is the limited reproducibility of the CNT growth process. An alternative approach to lateral integration is the manufacturing of arrays of CNTs based on vertical structures. Very homogeneous and reproducible growth of vertical CNT arrays by pyrolysis of acetylene on cobalt coated alumina substrates has been reported [394]. The structure and distribution of the diameters are shown in Fig. 9.22. The hexagonal cells, being open on top, have an average radius of 47 nm. They are positioned very symmetrically with an average distance of 98 nm. The manufacturing of vertical CNT transistors has already been reported [395]. However, they operated only at cryogenic temperatures of 4 K.

The first successful integration of CNT field effect transistors has been done using lateral structures similar to the above-shown TUBEFET but with other materials [396]. A structure using gold drain and source contacts is shown in Fig. 9.23. In this case, an aluminum gate contact has been covered by a 100 nm thick

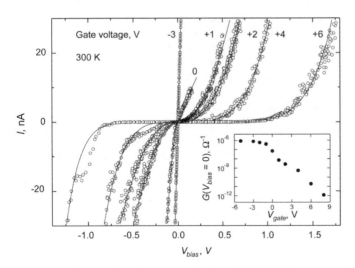

Fig. 9.21   Electrical device characteristics of a TUBEFET measured at 300 K. Inset: channel conductance *vs.* gate voltage (from [389])

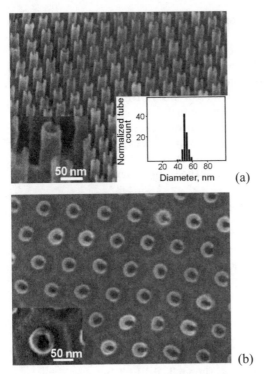

Normalized tube count

40
20

20 40 60 80
Diameter, nm   (a)

50 nm

50 nm   (b)

Fig. 9.22   SEM image and histogram of the diameters of the nanotubes of a vertical carbon nanotube array (from [394])

Al_2O_3 layer as gate isolator. Both layers have been deposited on top of a SiO_2 layer as substrate. The resulting enhancement-type p-channel MOSFETs with a voltage gain exceeding 10 have been wired together by gold metallization. In Fig. 9.24, the transfer characteristics of various digital functional circuits implemented by this technique are shown. Up to three interconnected transistors were used to implement an inverter, a NOR gate, a static RAM cell, and a ring-counter—with rather long switching times, however.

As a perspective for other applications of carbon nanotubes for electronic devices, it should be mentioned that heterojunctions between CNTs and silicon quantum wires have already been reported [397]. In this case, the silicon quantum wires were grown by CVD deposition in a silane atmosphere selectively on top of the CNTs. They consisted of a crystalline core covered by a thin amorphous silicon layer and a SiO_2 layer. The electrical characterization of this heterostructure showed a behavior similar to a Schottky diode and the current-voltage characteristics clearly exhibited rectifying behavior.

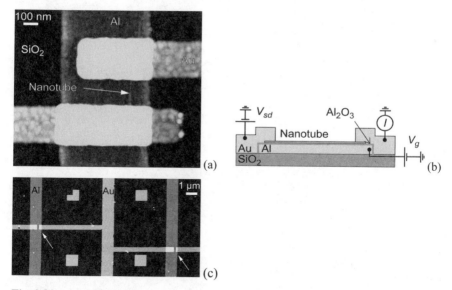

Fig. 9.23   (a) AFM image of a single SET, (b) schematic structure, and (c) AFM image of an integrated circuit structure using single wall carbon nanotube transistors (from [396])

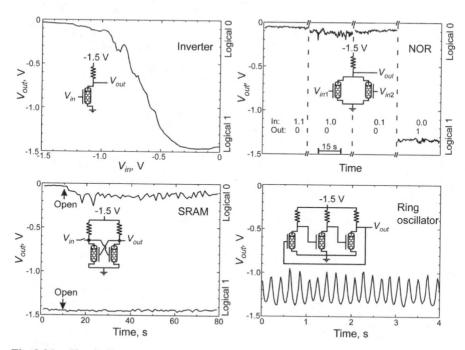

Fig. 9.24   Circuit diagrams and measured transfer characteristics of integrated circuits produced with carbon nanotube p-MOS transistors (from [396])

References

1   Historic Development

1. Semiconductor Industry Association (1999) The National Technology Roadmap for Semiconductors. San Jose (CA)
2. Drexler KE, Peterson C, Pergamit G (1991) Unbounding the future: the nanotechnology revolution. Addison-Wesley, New York

2   Quantum Mechanical Aspects

3. Clark T (1985) A Handbook of Computational Chemistry. Wiley & Sons, New York
4. Pople JA, Beveridge DL (1970) Approximate Molecular Orbital Theory. McGraw-Hill, New York
5. Deak P (1999) Approximate and parameterized quantum chemical methods for structural modeling of c-Si. In: Hull R (ed) Properties of Crystalline Silicon. Inspec, London, p 245
6. http://www.scientificamerican.com/exhibit/042897gear/042897nano.html (Status of 1997)
7. Ghadiri R (1996) Molecular Engineering. In: Crandall BC (ed) Nanotechnology. The MIT Press, Cambridge (MA)
8. Delley B, Steigmeier EF (1995) Size Dependency of Band Gaps in Silicon Nanostructures. Appl Phys Lett 67:2370
9. Maus M, Ganteför G, Eberhardt W (2000) The electronic structure and the band gap of nano-sized Si particles: competition between quantum confinement and surface reconstruction, Appl Phys A70:535
10. Kityk IV, Kassiba A, Tuesu K, Charpentier C, Ling Y, Makowska-Janusik M (2000) Vacancies in SiC nanopowders. Mat Sci Eng B77:147
11. Born M, Wolf E (1999) Principles of Optics: Electromagnetic Theory of Propagation, Interference and Diffraction of Light. Cambridge Univ Press, 7th ed, Oxford.
12. Liboff RL (1992) Introductory Quantum Mechanics. Addison-Wesley, 2nd ed, Reading
13. von Klitzing K, Dorda G, Pepper M (1980) New Method for High Accuracy Determination of Fine-Structure Constant Based on Quantized Hall Resistance. Phys Rev Lett, vol 45, p 494
14. Pasquarello A, Hybertsen MS, Car R (1998) Interface Structure between Silicon and its Oxide by First-principles Molecular Dynamics. Nature, vol 396, p 58

15.   Harrison RW (1999) Integrating Quantum and Molecular Dynamics. J Comp Chem, vol 20, p 1618
16.   Car R (1996) Modeling Materials by ab-initio Molecular Dynamics. Kluwer Acad Publ, Norwell
17.   Wirth GI (1999) Mesoscopic Phenomena in Nanometer Scale MOS Devices. Ph. D. Thesis, University of Dortmund, Germany

3   Nanodefects

18.   Hull R (1999) Properties of Crystalline Silicon. Inspec, London
19.   Pantelides ST (1986) Deep Centers in Semiconductors. Gordon and Breach, Newark (NJ)
20.   Dash WC (1957) Dislocations and Mechanical Properties of Crystals. In: Fisher JC *et al.* (eds). Wiley & Sons, New York
21.   Hwang KH, Park JW, Yoon E (1997) Amorphous {100} platelet formation in (100) Si induced by hydrogen plasma treatment. J Appl Phys, vol 81, p 74
22.   Job R, Ulyashin A, Fahrner WR (2000) The Evolution of Hydrogen Molecule Formation in Hydrogen Plasma Treated Czochralski Silicon. 2000 E-MRS Spring Meeting, Strasbourg, France
23.   Job R, Ulyashin A, Fahrner WR, Markevich VP, Murin LI, Lindström JL, Raiko V, Engemann J (2000) Bulk and Surface Properties of Cz-Silicon after Hydrogen Plasma Treatments. In: Claes CL *et al.* (eds) High Purity Silicon VI. El-Chem Soc Proc, vol 2000-17, p 209 (the 198th Meeting of the El-Chem Soc, Oct. 22–27, 2000, Phoenix)
24.   Kaufmann U, Schneider J (1976) Deep Traps in Semi-Insulating GaAs: Cr Revealed by Photosensitive ESR. Sol State Comm, vol 20, p 143
25.   Zerbst M (1966) Relaxationseffekte an Halbleiter-Isolator-Grenzflächen. Z Angew Phys, vol 22, p 30
26.   Klausmann E, Fahrner WR, Bräunig D (1989) The Electronic States of the Si-SiO$_2$ Interface. In: Barbottin G, Vapaille A (eds) Instabilities in Silicon Devices, vol 2. North-Holland, Amsterdam
27.   Ferretti R, Fahrner WR, Bräunig D (1979) High Sensitivity Non-Destructive Profiling of Radiation Induced Damage in MOS Structures. IEEE Trans Nucl Sci, NS-26, p 4828
28.   Schroder DK (1998) Semiconductor Material and Device Characterization. Wiley, New York
29.   van Wieringen A, Warmoltz N (1956) On the Permeation of Hydrogen and Helium in Single Crystal Silicon and Germanium at elevated Temperatures. Physica, vol 22, p 849
30.   Job R, Fahrner WR, Kazuchits NM, Ulyashin AG (1998) A Two Step Low Temperature Process for a p-n Junction Formation due to Hydrogen Enhanced Thermal Donor Formation in p-Type Czochralski Silicon. In: Nickel NH *et al.* (eds) Hydrogen in Semiconductors and Metals. MRS Symp Proc Ser, vol 513, p 337
31.   Ulyashin AG, Petlitskii AN, Job R, Fahrner WR (1998) Hydrogen Enhanced Thermal Donor Formation in p-Type Czochralski Silicon with Denuded Zone. In: Claeys CL *et al.* (eds) High Purity Silicon V. El-Chem Soc Proc, vol 98-13, p 425

32. Ulyashin AG, Ivanov AI, Job R, Fahrner WR, Frantskevich AV, Komarov FF, Kamyshan AC (2000) The hydrogen gettering at post-implantation plasma treatments of helium- and hydrogen implanted Czochralski silicon. Mat Sci Eng, vol B73, p 64

33. Rebohle L, von Borany J, Yankov RA, Skorupa W (1977) Strong blue and violet photoluminescence and electroluminescence from germanium-implanted and silicon-implanted silicon-dioxide layers. Appl Phys Lett, vol 71, p 2809

34. Rebohle L, von Borany J, Fröb H, Skorupa W (2000) Blue photo- and electroluminescence of silicon dioxide layers implanted with group IV elements. Appl Phys, vol B 71, p 131

35. Gebel T, Skorupa W, von Borany J, Borchert D, Fahrner WR (2000) Integrierter Optokoppler und Verfahren seiner Herstellung. Deutsches Patent 100 11 28.7 (March 8, 2000)

36. Thomas DF (2000) Porous Silicon. In: Nalwa HS (ed) Handbook of Nanostructured Materials and Nanotechnology, vol 4. Academic Press, New York

37. Bondarenko VP, Novikov AP, Shiryaev Yu C, Samoiluk TT, Timofeev AB (1987) A method for porous silicon production on silicon substrate (in Russian). Sowjetunion Patent N 1403902.

38. Koshida N, Koyama H (1992) Visible electroluminescence from porous silicon. Appl Phys Lett, vol 60, p 347

39. Asmus T, Fink D, Sieber I, Hoffmann V, Müller M, Stolterfoht N, Berdinsky AS. Deposition of Conducting Polymers into Ion Tracks. Unpublished.

40. Vobecky J, Hazdra P, Galster N, Carroll E (1998) Free-Wheeling Diodes with Improved Reverse Recovery by Combined Electron and Proton Irradiation. Proc 8th PECM, Prague, Czech Republic, Sep. 8–10, 1998

41. Schwuttke GH (1973) Damage Profiles in Silicon and Their Impact on Device Reliability. Tech Rep 3, ARPA contract DAHC 15-72-C-0274

42. Bergmann RB, Rinke TJ (2000) Perspective of Crystalline Si Thin Film Solar Cells: A New Era of Thin Monocrystalline Si Films? Prog Photovolt: Res Appl, vol 8, p 451

43. Denisenko A, Aleksov A, Pribil A, Gluche P, Ebert W, Kohn E (2000) Hypothesis on the Conductivity Mechanism in Hydrogen Terminated Diamond Films. Diam Rel Mat, vol 9, p 1138

44. Ulyashin AG, Gelfand RB, Shopak NV, Zaitsev AM, Denisenko AV, Melnikov AA (1993) Passivation of Boron Acceptor in Diamond by Atomic Hydrogen: Molecular-orbital Linear-combination-of-atomic-orbitals Simulation and Experimental Data. Diam Rel Mat, vol 2, p 1516

45. Martin CR (1994) Nanomaterials: A Membrane-based Synthetic Approach. Science, vol 266, p 1961

46. Biswas A, Awasthi DK, Singh BK, Lotha S, Singh JP, Fink D, Yadav BK, Bhattacharya B, Bose SK (1999) Resonant Electron Tunnelling in Single Quantum Well Heterostructure Junction of Electrodeposited Metal Semiconductor Nanostructures Using Nuclear Track Filters. Nucl Instr Meth B 151, p 84

47. Yoshida M, Tamada K, Spohr R. Pers comm.

48. Granstöm M, Berggren M, Inganäs O (1995) Micrometer-and nanometer-sized polymeric light-emitting diodes. Science, vol 267, p 1479

49. Fink D, Schulz A, Müller M, Richter H, Danziger M. Ion-Track Based Microinductivities. Unpublished.

50.   Zorinants G, Fink D (1997) Unpublished.
51.   Berdinsky S, Fink D. Unpublished.
52.   Könenkamp R. Unpublished.

4   Nanolayers

53.   Choy KL (2000) Vapor Processing of Nanostructured Materials. In: Nalwa HS (ed) Handbook of Nanostructured Material and Nanotechnology. Academic Press, New York
54.   Graper EB (1995) Resistance Evaporation. In: Glocker AD, Shah SI (eds) Handbook of Thin Film Process Technology. Inst of Phys, Bristol
55.   Graper EB (1995) Electron Beam Evaporation. In: Glocker AD, Shah SI (eds) Handbook of Thin Film Process Technology. Inst of Phys, Bristol
56.   Shah SI (1995) Sputtering: Introduction and General Discussion. In: Glocker AD, Shah SI (eds) Handbook of Thin Film Process Technology. Inst of Phys, Bristol
57.   Graper EB (1995) Ion Vapour Deposition. In: Glocker AD, Shah SI (eds) Handbook of Thin Film Process Technology. Inst of Phys, Bristol
58.   Kawasaki M, Gong J, Nantoh M, Hasegawa T, Kitazawa K, Kumagai M, Hirai K, Horiguchi K, Yoshimoto M, Koinuma H (1993) Preparation and Nanoscale Characterization of Highly Stable YBa$_2$Cu$_3$O$_{7-\delta}$ Thin Films. Jpn J Appl Phys, vol 32, p 1612
59.   Tsang WT (1985) Molecular Beam Epitaxy for III-V Compound Semiconductors. In: Willardson RK, Beer RC (eds) Semiconductors and Semimetals, vol 22, part A. Academic Press, New York, p 96
60.   Joyce BA, Foxton CT (1977) Growth and Doping of Semiconductor Films by Molecular Beam Epitaxy. Solid State Device (1976), ESSDERC Sep 13–16, 1976, Inst Phys Conf No 32
61.   Hansen M (1958) Constitution of Binary Alloys. McGraw-Hill, New York
62.   Wagemann HG. Pers comm.
63.   Webber RF, Thorn RS, Large LN (1969) The Measurement of Electrical Activity and Hall Mobility of Boron and Phosphorus Ion-implanted Layers in Silicon. Int J Electronics, vol 26, p 163
64.   Dearnaley G, Freeman JH, Nelson RS, Stephen J (1973) Ion Implantation. North Holland, Amsterdam
65.   Gibbons JF, Johnson WS, Mylroie SW (1975) Projected Range Statistics. Dowden, Hutchinson, and Ross, Stroudsburg (PA)
66.   Maxwell jr HR (1985) Process Data. In: Beadle WE, Tsai JC, Plummer RD (eds) Quick Reference Manual for Silicon Integrated Circuit Technology. Wiley-Interscience, New York
67.   Rappich J (2002) Niedertemperatur-Passivierung. http://www.hmi.de/bereiche/SE/SE1/projekte/t_verfahren/zelltechnologie/niedertemp/index.html   (Status of Nov. 2002)
68.   Pliskin WA, Zanin SJ (1970) Film Thickness and Composition. In: Glang R, Maissel LI (eds) Handbook of Thin Film Technology. p 11.6, McGraw-Hill, New York
69.   Dorenwendt K (1985) Interferometrie. In: Kohlrausch F (ed) Praktische Physik.

23. ed, p 670, Teubner-Verlag, Stuttgart, Germany

70. Sugawara K, Nahazawa Y, Yoshimi T (1976) In Situ Thickness Monitoring of Thick Polycrystalline Silicon Film and Its Application to Silicon Epitaxial Growth. J El Chem Soc, vol 123, no 4, p 586

71. Archer RJ (1962) Determination of the Properties of Films on Silicon by Method of Ellipsometry. J Opt Soc Am, vol 52, p 970

72. Grabosch G, Fahrner WR (2000) Spectral Ellipsometry and Dark Conductivity Measurements on p- and n-type Microcrystalline Films. Micromat 2000, Apr. 17–19, Berlin, Germany

73. Pliskin WA, Zanin SJ (1970) Film Thickness and Composition. In: Glang R, Maissel LI (eds) Handbook of Thin Film Technology. p 11.30, McGraw-Hill, New York

74. Fries T. Pers comm.

75. Weima JA, Job R, Fahrner WR, Kosaca G, Müller N, Fries T (2001) Surface analysis of ultra-precisely polished chemical vapor deposited diamond films using spectroscopic and microscopic techniques. J Appl Phys, vol 89, p 2434

76. Schwuttke GH (1974) Damage Profiles in Silicon and Their Impact on Device Reliability. Technical Report No. 5, ARPA contract DAHC15-72-C-0274

77. Schwuttke GH (1965) New X-Ray Diffraction Microscopy Technique for the Study of Imperfections in Semiconductor Crystals. J Appl Phys, vol 36, p 2712

78. Chang SL, Thiel PA (1995) Low-Energy Electron Diffraction. In: Glocker AD, Shah SI (eds) Handbook of Thin Film Process Technology. Inst of Phys, Bristol

79. Taylor NJ (1966) A LEED Study of the Epitaxial Growth of Copper on the (110) Surface of Tungsten. Surf Sci, vol 4, p 161
    Photographs compiled by Khan IH (1970). In: Maissel RI, Glang R (eds) Handbook of Thin Film Technology, p 10-1. McGraw-Hill, New York

80. Joyce BA (1995) Reflection High-Energy Electron Diffraction as a Diagnostic Technique in Thin Film Growth Studies. In: Glocker AD, Shah SI (eds) Handbook of Thin Film Process Technology. Inst of Phys, Bristol

81. Cain OJ, Vook RW (1978) Epitaxial Layers of Cu_2S Grown from Liquid Solution and Investigated by RHEED. J Electrochem Soc, vol 125, p 882

82. Schindler R (1996) Semiconductor Technology. Skripte der FernUniversität Hagen

83. Grasserbauer M, Dudek HJ, Ebel MF (1985) Angewandte Oberflächenanalyse. Springer-Verlag, Berlin

84. Job R, Ulyashin AG, Fahrner WR, Ivanov AI, Palmetshofer L (2001) Oxygen and hydrogen accumulation at buried implantation-damage layers in hydrogen- and helium-implanted Czochralski silicon. Appl Phys, vol A 72, p 325

85. Baek SK, Choi CJ, Seong TY, Hwang H, Kim HK, Moon DW (2000) Characterization of sub-30 nm p^+/n Junction Formed by Plasma Ion Implantation. J Electrochem Soc, vol 147, p 3091

86. Lifshin E (1994) Electron Microprobe Analysis. In: Cahn RW, Haasen P, Kramer EJ (eds) Material Science and Technology, vol 2B. VCH, Weinheim

87. Physical Electronics Industries (1972) Untitled.

88. Pawlik D, Oppolzer H, Hilmer T (1985) Characterization of thermal oxides grown on $TaSi_2$/polysilicon films. J Vac Sci Technol, vol B3, p 492

89. Irvin JC (1962) Resistivity of Bulk Silicon and of Diffused Layers. BSTJ, vol 41, p 387

90.   van der Pauw LJ (1958) A Method of Measuring Specific Resistivity and Hall Effect of Discs of Arbitrary Shape. Phil Res Rep, vol 13, p 1

91.   Fahrner WR, Klausmann E, Bräunig D (1987) Si/SiO$_2$ Intrinsic States and Interface Charges. Scientific Report of the Hahn-Meitner Institute

92.   Fahrner WR, Bräunig D, Knoll M, Laschinski JR (1984) Ion Implantation for Deep (>100 μm) Buried Layers. In: Gupta DC (ed) Semiconductor Processing, ASTM STP 850. American Society for Testing and Materials

93.   Schreiber T (2001) Untitled. Materials, vol 13, p 11. Unaxis Semiconductors, Balzers, Liechtenstein

94.   Pliskin WA, Conrad EE (1964) Nondestructive Determination of Thickness and Refractive Index of Transparent Films. IBM J Res Develop, vol 8, p 43

95.   Pliskin WA, Resch RP (1965) Refractive Index of SiO$_2$ Films Grown on Silicon. J Appl Phys, vol 36, p 2011

96.   Reizman F, van Gelder WE (1967) Optical Thickness Measurements of SiO$_2$-Si$_3$N$_4$ Films on Si. Sol-State Electr, vol 10, p 625

97.   Runyan WR (1965) Silicon Semiconductor Technology. McGraw-Hill, New York

98.   Borchert D, Wolffersdorf C, Fahrner WR (1995) A Simple Compact Measurement Set-up for the Optical Characterization of Solar Cells. 13th Europ Photovolt Solar Energy Conf, Nice, France, Oct. 23–27, 1995

99.   Blaustein P, Hahn S (1989) Realtime Inspection of Wafer Surfaces. Sol State Technol, vol 32, no 12, p 27

100.   Jahns J (1998) Free-Space Optical Digital Computing and Interconnection. Prog Opt, p 419, Wolf E (ed), Elsevier, Amsterdam

101.   Fey D, Erhard W, Gruber M, Jahns J, Bartelt H, Grimm G, Hoppe L, Sinzinger S (2000) Optical Interconnects for Neural and Reconfigurable VLSI Architectures. Proc IEEE, vol 88, p 838

5   Nanoparticles

102.   Jiang J, Lau M, Tellkamp VL, Lavernia EJ (2000) Synthesis of Nanostructured Coatings by High-Velocity Oxygen-Fuel Thermal Spraying. In: Nalwa HS (ed) Handbook of Nanostructured Materials and Nanotechnology, vol 1, p 159. Academic Press, New York

103.   Siegel RW (1993) Synthesis and properties of nanophase materials. Mat Sci Eng A, vol 168, p 189

104.   Grandjean N, Massies J (1993) Epitaxial Growth of highly strained In$_x$Ga$_{1-x}$ on GaAs(001): the role of surface diffusion length. J Cryst Growth, vol 134, p 51

105.   Nötzel R (1996) Self-organized growth of quantum-dot structures. Semicond Sci Technol, vol 11, p 1365

106.   Perrin J, Aarts, JF (1983) Dissociative Excitation of SiH$_4$, SiD$_4$, Si$_2$H$_6$ and GeH$_4$ by 0–100 eV electron impact. Chem Phys, vol 80, p 351

107.   Sobolev VV, Guilemany JM, Calero JA (1995) Dynamic Processes during In-Flight Motion of Cr$_3$C$_2$-NiCr Powder Particles in High Velocity Oxy-Fuel (HVOF) Spraying. J Mater Process Manuf Sci, vol 4, p 25

108.   McClelland JJ (2000) Nanofabrication via Atom Optics. In: Nalwa HS (ed) Hand-

book of Nanostructured Materials and Nanotechnology, vol 1, p 335. Academic Press, New York

109.   McClelland JJ, Gupta R, Jabbour ZJ, Celotta RJ (1966) Laser Focusing of Atoms for Nanostructure Fabrication. Aust J Phys, vol 49, p 555

110.   Gupta R, McClelland JJ, Jabbour ZJ, Celotta RJ (1995) Nanofabrication of a two-dimensional array using laser focused atomic deposition. Appl Phys Lett, vol 67, p 1378

111.   Kwiatkowski KC, Lukehart CM (2000) Nanocomposites Prepared by Sol-Gel Methods: Synthesis and Characterization. In: Nalwa HS (ed) Handbook of Nanostructured Materials and Nanotechnology, vol 1, p 387. Academic Press, New York

112.   Xie Y, Qian Y, Wang W, Zhang S, Zhang Y (1996) A Benzene-Thermal Synthetic Route to Nanocrystalline GaN. Science, vol 272, p 1926

113.   Janik JF, Wells RL (1996) Gallium Imide, {Ga(NH)$_{3/2}$}$_n$, a New Polymeric Precursor for Gallium Nitride Powders. Chem Mater, vol 8, p 2708

114.   Ozaki N, Ohno Y, Takeda S (1999) Optical properties of Si nanowires on a Si{111} surface. Mat Res Symp Proc, vol 588, p 99, Ünlü MS, Piqueras J, Kalkhoran NM, Sekigushi T (eds), Warrendale (PA)

115.   Teng CW, Muth JF, Kolbas RM, Hassan KM, Sharma AK, Kvit A, Narayan J (1999) Quantum Confinement of above-Band-Gap Transitions in Ge Quantum Dots. Mat Res Symp Proc, vol 588, p 263, Ünlü MS, Piqueras J, Kalkhoran NM, Sekigushi T (eds), Warrendale (PA)

116.   Mittleman DM, Schoenlein RW, Shiang JJ, Colvin VL, Alivisatos AP, Shank CV (1994) Quantum size dependence of femtosecond electronic dephasing and vibrational dynamics in CdSe nanocrystals. Phys Rev, vol B49, p 14435

117.   Cao X, Koltypin Y, Kataby G, Prozorov R, Gedanken A (1995) Controlling the particle size of amorphous iron nanoparticles. J Mat Res, vol 10, p 2952

118.   Simon U, Schön G (2000) Electrical Properties of Chemically Tailored Nanoparticles and Their Application in Microelectronics. In: Nalwa HS (ed) Handbook of Nanostructured Materials and Nanotechnology, vol 3, p 131. Academic Press, New York

119.   http://www.nanonet.de/anw-01.htm (Status of May 2001)

120.   Choi WB, Chung DS, Kang JH, Kim HY, Jin YW, Han IT, Lee YH, Jung JE, Lee NS, Park GS, Kim JM (1999) Fully sealed, high-brightness carbon-nanotube field emission display. Appl Phys Lett, vol 75, p 3129

6   Selected Solid States with Nanocrystalline Structures

121.   Grabosch, G (2000) Herstellung und Charakterisierung von PECVD abgeschiedenem mikrokristallinem Silizium. Dissertation am Fachbereich Elektrotechnik der FernUniversität Hagen.

122.   Hattori Y, Kruangam D, Katoh K, Nitta Y, Okamoto H, Hamakawa Y (1987) High-Conductive Wide Band Gap p-Type a-SiC:H Prepared by ECR CVD and Its Application to High Efficiency a-Si Basis Solar Cells. Proc 19th IEEE Photovolt Spec Conf, p 689

123. Konaga M, Takai H, Kim WY, Takahashi K (1985) Preparation of Amorphous Silicon and Related Semiconductors by Photochemical Vapor Deposition and Their Application to Solar Cells. Proc 19th IEEE Photovolt Spec Conf, p 1372

124. Kuwanu Y, Tsuda S (1998) High Quality p-Type a-SiC Film Doped with B(CH$_3$)$_3$ and its Application to a-Si Solar Cells. Mat Res Soc Symp Proc, vol 118, p 557, Mat Res Soc

125. Guha S, Ovshinsky SR (1988) P and N-Type Microcrystalline Semiconductor Alloy Material Including Band Gap Widening Elements, Devices Utilizing Same. U.S. Patent No. 4,775,425

126. Dusane SR (1992) Gap states in hydrogenated microcrystalline silicon by glow discharge technique. J Appl Phys, vol 72, p 2923

127. Lucovsky G, Wang C, Chen YL (1992) Barrier-limited transport in μc-Si and μc-SiC thin Films prepared by remote plasma-enhanced chemical vapor deposition. J Vac Sci Technol A, vol 10, p 2025

128. Shimizu I, Hanna HI, Shirai H (1990) Control of Chemical Reactions for Growth of Crystalline Si at Low Substrate Temperature. Mat Res Soc Symp Proc, vol 164, p 195, Mat Res Soc

129. Komiya T, Kamo A, Kujirai H, Shimizu I, Hanna JI (1990) Preparation of Crystalline Si Thin Films by Spontaneous Chemical Deposistion. Mat Res Soc Symp Proc, vol 164, p 63, Mat Res Soc

130. Tonouchi M, Moriyama F, Miyasato T (1990) Characterization of μc-Si:H Films Prepared by H$_2$ Sputtering. Jap J Appl Phys, vol 29, p L385

131. Feng GF, Katiyar M, Yang YH, Abelson JR, Maley N (1992) Growth and Structure of Microcrystalline Silicon by Reactive DC Magnetron Sputtering. Mat Res Soc Symp Proc, vol 258, p 179, Mat Res Soc

132. Chaudhuri P, Ray S, Barua AK (1984) The Effect of Mixing Hydrogen with Silane on the Electronic and Optical Properties of Hydrogenated Amorphous Silicon Films. Thin Solid Films, vol 113, p 261

133. Shirafuji J, Nagata S, Kuwagaki M (1985) Effect of hydrogen dilution of silane on optoelectronic properties in glow discharged hydrogenated silicon films. J Appl Phys, vol 58, p 3661

134. Meiling H, van den Boogard MJ, Schropp RE, Bezemer J, van der Weg WF (1990) Hydrogen Dilution of Silane: Correlation between the Structure and Optical Band Gap in GD a-Si:H Films. Mat Res Soc Symp Proc, vol 192, p 645, Mat Res Soc

135. Hollingsworth RE, Bhat PK, Madan A (1987) Microcrystalline and Wide Band Gap p^{+} Window Layers for a-Si p-i-n Solar Cells. J Non-Cryst Sol, vol 97 & 98, p 309

136. Kroll U, Meier J, Torres P, Pohl J, Shah A (1998) From amorphous to microcrystalline silicon films prepared by hydrogen dilution using VHF (70 MHz) GD technique. J Non-Cryst Sol, vol 227–230, p 69

137. Tsai CC, Thompson R, Doland C, Ponce FA, Anderson GB, Wacker B (1998) Transition from Amorphous to Crystalline Silicon: effect of Hydrogen on Film Growth. Mat Res Soc Symp Proc, vol 118, p 49, Mat Res Soc

138. Schropp RE, Zeman M (1998) Amorphous and Microcrystalline Silicon Solar Cells: Modelling, Materials and Devices Technology. Kluwer Acad. Publ, Boston (MA)

139. Matz W. Pers comm.

140.   Klug HP, Alexander LE (1974) X-Ray Diffraction Procedures. Wiley & Sons, 2nd ed, New York

141.   Borchert D, Hussein R, Fahrner WR (1999) A Simple (n) a-Si / (p) c-Si Hetero-junction Cell Process Yielding Conversion Efficiencies up to 15.3 %. 11th Int Photovolt Sci Eng Conf, Sapporo, Japan

142.   Kawamoto K, Nakai T, Bab T, Taguchi M, Sakata H, Tsuge S, Uchihashi K, Tanaka M, Kiyama S (2001) A High Efficiency HIT Solar Cell (21.0 % $\approx$ 100 cm^2) with Excellent Interface Properties. 12th Int Photovolt Sci Eng Conf, Jeju, Korea

143.   Barrer RM (1978) Zeolites and Clay Minerals as Sorbents and Molecular Sieves. Academic Press, London

144.   Derouane EG, Lemos F, Naccache C, Ribeiro FR (1992) Zeolite Microporous Solids: Synthesis, Structure, and Reactivity. Kluwer Academic Publishers, Dordrecht

145.   Zhen S, Seff K (2000) Structures of organic sorption complexes of zeolites. Microporous and Mesoporous Materials, vol 39, p 1

146.   Caro J, Noack M, Kolsch P, Schafer R (2000) Zeolite membranes – state of their development and perspective. Microporous and Mesoporous Materials, vol 38, p 3

147.   Pauling L (1976) Die Natur der chemischen Bindung. Verlag Chemie, Weinheim

148.   Pauling L (1988) General Chemistry. Dover Publications Inc, New York

149.   Meier WM, Olson DH, Baerlocher C (1996) Atlas of Zeolite Structure Types. Zeolites, vol 17, p 1

150.   IZA Structure Commission. Database of Zeolite Structures. http://www.zeolites.ethz.ch/Zeolites/Introduction.htm (Status of Nov. 2002)

151.   Robson H (1998) Verified Synthesis of Zeolite Materials. Microporous Materials, vol 22, p 495

152.   Barrer RM (1981) Hydrothermal Chemistry of Zeolites. Academic Press, London

153.   Cheetham AK, Férey G, Loiseau T (1999) Anorganische Materialien mit offenen Gerüsten. Ang Chem, vol 111, p 3466

154.   Charnell JF (1971) Gel Growth of Large Crystals of Sodium and Sodium X Zeolites. J Cryst Growth, vol 8, p 291

155.   Shimizu S, Hamada H (1999) Synthese riesiger Zeolithkristalle durch langsame Auflösung kompakter Ausgangsmaterialien. Angew Chem, vol 111, p 2891

156.   McCusker LB (1998) Product characterization by X-ray powder diffraction. In: Robson H (ed) Verified Synthesis of Zeolite Materials. Microporous Materials, vol 22, p 527

157.   Jansen K (1998) Characterization of zeolites by scanning electron microscopy. In: Robson H (ed) Verified Synthesis of Zeolite Materials. Microporous Materials, vol 22, p 531

158.   Stöcker M (1998) Product characterization by nuclear magnetic resonance. In: Robson H (ed) Verified Synthesis of Zeolite Materials. Microporous Materials, vol 22, p 533

159.   Kuzmany H (1998) Solid-State Spectroscopy. Springer, Berlin

160.   Ruthven DM (1998) Characterization of zeolites by sorption capacity measurements. In: Robson H (ed) Verified Synthesis of Zeolite Materials. Microporous Materials, vol 22, p 537

161.   Dyer A (1998) Ion-exchange capacity. In: Robson H (ed) Verified Synthesis of

Zeolite Materials. Microporous Materials, vol 22, p 543

162. Karge HG (1998) Characterization by infrared spectroscopy. In: Robson H (ed) Verified Synthesis of Zeolite Materials. Microporous Materials, vol 22, p 547

163. Flanigen EM, Sands LB (1998) Advances in Chemistry Series 101. American Chemical Society, Washington DC, p 201

164. Knops-Gerrits PP, De Vos DE, Feijen EJ, Jacobs PA (1997) Raman spectroscopy of zeolites. Microporous Materials, vol 8, p 3

165. Breck DW (1974) Zeolithe Molecular Sieves: Structure, Chemistry and Use. Wiley, New York

166. Schmid G (1994) Clusters and Colloids. Wiley-VCH, Weinheim

167. Gonsalves KE, Rangarajan, Wang J (2000) Chemical Synthesis of Nanostructured Metals, Metal Alloys, and Semiconductors. In: Nalwa HS (ed) Handbook of Nanostructured Materials and Nanotechnology. Academic Press, London

168. Simon U, Franke ME (2000) Electrical Properties of nanoscaled host/ guest compounds. Microporous and Mesoporous Materials, vol 41, p 1

169. Exner D, Jäger NI, Kleine A, Schulz-Ekloff G (1998) Reduction-Agglomeration Model for Metal Dispersion in Platinum-exchanged NaX Zeolite. J Chem Faraday Trans, vol 84, p 4097

170. Ozin GA, Kuperman A, Stein A (1989) Advanced Zeolite Material Science. Angew Chem Intern Ed Engl, vol 28, p 359

171. Wang Y, Herron N, Mahler W, Suna H (1989) Linear- and nonlinear-optical properties of semiconductor clusters. J Opt Soc Am, vol B6, p 808

172. Blatter F, Blazey KW (1990) Conduction-electron resonance of Na-Cs alloys in zeolite Y. IBM Research Report, Zürich

173. Haug K, Srdanov VI, Stucky GD, Metiu H (1992) The absorption spectrum of an electron solvated in sodalite. J Chem Phys, vol 96, p 3495

174. Ozin GA (1992) Nanochemistry: Synthesis in Diminishing Dimensions. Adv Mater, vol 10, p 612

175. Ozin GA, Özkar S (1992) Zeolites: A Coordination Chemistry View of Metal-Ligand Bonding in Zeolite Guest-Host Inclusion Compounds. Chem Mater, vol 4, p 551

176. Edwards PP, Woodall LJ, Anderson PA, Armstrong AR, Slaski M (1993) On the Possibility of an Insulator-Metal Transition in Alkali Metal Doped Zeolites. Chem Soc Rev, vol 22, p 305

177. Bowes CL, Ozin GA (1998) Tin sulfide clusters in zeolite A, Sn_4S_6-Y. J Mat Chem, vol 8, p 1281

178. Herron N (1986) A Cobalt Oxygen Carrier in Zeolite Y. A Molecular 'Ship in a Bottle'. Inorg Chem, vol 25, p 4714

179. Herron N, Wang Y, Eddy M, Stucky GD, Cox DE, Bein T, Moller K (1989) Structure and Optical Properties of CdS Superclusters in Zeolite Hosts. J Am Chem Soc, vol 111, p 530

180. Hirono T, Kawana A, Yamada T (1987) Photoinduced effects in a mordenite-AgI inclusion compound. J Appl Phys, vol 62, p 1984

181. Nozue Y, Tang ZK, Goto T (1990) Excitions in PbI_2 Clusters Incorporated into Zeolite Cages. Solid State Commun, vol 73, p 31

182. Liu X, Thomas JK (1989) Formation and Photophysical Properties of CdS in Zeolites with Cages and Channels. Langmuir, vol 5, p 58

183. Schwenn HJ, Wark M, Schulz-Ekloff G, Wiggers H, Simon U (1997) Electrical

and optical properties of zeolite Y supported SnO_2 nanoparticles. Colloid Polym Sci, vol 275, p 91

184.   Wark M, Schwenn HJ, Schulz-Ekloff G, Jäger NI (1992) Structure, Photoabsorption and Reversibles Reactivity of Faujasite-Supported Dispersions of CdO and SnO_2. Ber Bunsenges Phys Chem, vol 96, p 1727

185.   Brus LE (1986) Electronic wave functions in semiconductor clusters: experiment and theory. J Chem Phys, vol 90, p 2555

186.   Anderson PA, Bell RG, Catlow SR, Chang FL, Dent AJ, Edwards PP, Gameson I, Hussain I, Porch A, Thomas JM (1996) Matrix-Bound Nanochemical Possibilities. Chem Mater, vol 8, p 2114

187.   Kelly MJ (1995) A Model Electronic-Structure for Metal Intercalated Zeolites. J Phys Condensed Matter, vol 7, p 5507

188.   Anderson PA, Edwards PP (1992) A Magnetic Resonance Study of the Inclusion Compounds of Sodium in Zeolites: Beyond the Metal Particles. J Am Chem Soc, vol 114, p 10608

189.   Armstrong AR, Anderson PA, Edwards PP (1994) The Composition Dependence of the Structure of Potassium-Loaded Zeolite-A – Evolution of a Potasium Superlatice. J Solid State Chem, vol 111, p 178

190.   Anderson PA, Armstrong AR, Edwards PP (1994) Ionization and Delocalization in Potassium Zeolite-L – A Combined Neutron-Diffraction and Electron-Spin-Resonance Study. Angew Chem Intern Ed. Engl, vol 33, p 641

191.   Pan M (1996) High Resolution Electron Microscopy in Zeolites. Micron, vol 7, p 219

192.   Gallezot P (1979) The state and catalytic properties properties of platinum and palladium in faujasite-type zeolites. Catal Rev Sci Eng, vol 20, p 121

193.   Homeyer ST, Sachtler WM (1989) Elementary steps in the formation of highly dispersed palladium in NaY. J Catal, vol 118, p 266

194.   Stein A, Ozin GA (1993) Sodalite Superlattices: from molecules to clusters to expanded insulators, semiconductors and metals. In: von Ballmoos R, Higgins JB, Treacy MM (eds) Proc 9th Intern Zeolites Conf, vol I, p 93, Butterworth-Heinemann

195.   Aparisi A, Fornés V, Márquez F, Moreno R, López C, Meseguer F (1996) Synthesis and optical properties of CdS and Ge clusters in zeolite cages. Solid State Electronics, vol 40, p 641

196.   Deson J, Lalo C, Gédéon A, Vasseur F, Fraissard J (1996) Laser-induced luminescence in reduced copper-exchanged Y zeolite. Chemical Physics Letters, vol 258, p 381

197.   Chen W, Wang Z, Lin L, Lin J, Su M (1997) Photostimulated luminescence of silver clusters in zeolite-Y. Physics Letters, vol A 232, p 391

198.   Chen W, Wang Z, Lin Z, Lin L, Fang K, Xu Y, Su M, Lin J (1998) Photostimulated luminescence of AgI clusters in zeolite-Y. J Appl Phys, vol 83, p 3811

199.   Armand P, Saboungi ML, Price DL, Iton L, Cramer C, Grimsditch M (1997) Nanoclusters in Zeolite. Phys Rev Lett, vol 79, p 2061

200.   Mitsa V, Fejsa I (1997) Raman spectra of chalcogenides implanted into pores of zeolites. J Molecular Structure, vol 410–411, p 263

201.   Price GL, Kanazirev V (1997) Guest ordering in zeolite hosts. Zeolites, vol 18, p 33

202.   Caro J, Noack M, Kolsch P, Schafer R (2000) Zeolite membranes – state of their

development and perspective. Microporous and Mesoporous Materials, vol 38, p 3

203. Walcarius A (1999) Zeolite-modified electrodes in electrochemical chemistry. Analytica Chimica Acta, vol 384, p 1

204. Weitkamp J, Fritz M, Ernst S (1995) Zeolites as media for hydrogen storage. Int J Hydrogen Energy, vol 20, p 967

205. Kynast U, Weiler V (1994) Efficient luminescence from zeolites. Adv Mater, vol 6, p 937

206. Kelly MJ (1993) The poor prospects for one-dimensional devices. Int J Electron, vol 75, p 27

207. Videnova-Adrabinska V (1995) The hydrogen bond as a design element in crystal engineering. Two- and three-dimensional building blocks of crystal architecture. J Molecular Structures, vol 374, p 199

208. Edwards PP, Anderson PA, Woodall LJ, Porch A, Armstrong AR (1996) Can we synthesise a dense bundle of quasi one-dimensional metallic wires? Mater Sci Engineer, vol A 217/218, p 198

209. Anderson PA, Armstrong AR, Edwards PP (1994) Ionisierung und Elektronende-lokalisierung in Kalium-Zeolith-L: eine kombinierte Neutronenbeugungs- und ESR-Studie. Angew Chem, vol 106, p 669

210. Rabo JA, Angell CL, Kasai PH, Schomaker V (1996) Studies of Cations in Zeo-lite: Adsorption of Carbon Monoxide; Formation of Ni ions and Na_{3+4} centres. Disc Faraday Soc, vol 41, p 328

211. Edwards PP, Harrison MR, Klinowski J, Ramdas S, Thomas JM, Johnson DC, Page CJ (1984) Ionic and Metallic Clusters in Zeolite. J Chem Soc, Chem Commun, p 982

212. Harrison MR, Edwards PP, Klinowski J, Thomas JM, Johnson DC, Page CJ (1984) Ionic and Metallic Clusters of the Alkali Metals in Zeolite Y. J Solid State Chem, vol 54, p 330

213. Anderson PA, Singer RJ, Edwards PP (1991) A New Potassium Cluster in Zeo-lites X and A. J Chem Soc, Chem Commun, p 914

214. Anderson PA, Edwards PP (1992) A Magnetic Resonance Study in the Inclusion Compounds of Sodium in Zeolites: Beyond the Metal Particle Model. J Am Chem Soc, vol 114, p 10608

215. Armstrong AR, Anderson PA, Woodall LJ, Edwards PP (1994) Structure and Electronic Properties of Cesium-Loaded Zeolite A. J Phys Chem, vol 98, p 9279

216. Anderson PA, Edwards PP (1994) Reassessment of the conduction-electron spin resonance of alkali metals in zeolites. Phys Rev, vol B50, p 7155

7   Nanostructuring

217. Zaitsev AM, Kosaca G, Richarz B, Raiko V, Job R, Fries T, Fahrner WR (1998) Thermochemical polishing of CVD diamond films. Diam Rel Mat, vol 7, p 1108

218. Weima JA, Zaitsev AM, Job R, Kosaca GC, Blum F, Grabosch G, Fahrner WR (1999) Nano-Polishing and Subsequent Optical Characterization of CVD Poly-crystalline Diamond Films. Proc 25th Ann Conf IEEE Ind Elect Soc, p 50. IECON, San Jose (CA)

219. Weima JA, Fahrner WR, Job R (2001) Experimental investigation of the parameter dependency of the removal rate of thermochemically polished CVD diamonds. J Electrochem Soc, accepted for publication

220. Weima JA, Fahrner WR, Job, R (2001) A Model of the Thermochemical Polishing of CVD Diamond Films on Transition Metals with Emphasis on Steel. Submitted to the J Electrochem Soc

221. Weima JA, Job R, Fahrner WR (2002) Thermochemical Beveling of CVD Diamond Films Intended for Precision Cutting and Measurement applications. Diamond Relat Mat, vol 11, p 1537

222. Momose HS, Ono M, Yoshitomi T, Ohguro T, Nakamura S, Saito M Iwai H (1996) 1.5 nm Direct-Tunneling Gate Oxide Si MOSFET's. IEEE Trans, vol ED 43, p 1233

223. Hilleringmann U (1999) Silizium-Halbleitertechnologie. Teubner, Stuttgart, p 70

224. Fa. Oxford Instruments/Plasma Technology (Status of Sep. 2001) www.oxfordplasma.de

225. Cullmann E, Cooper K, Reyerse C (1991) Optimized Contact/Proximity Lithography. Suss Report, vol 5, p 1–4

226. Goodberlet JG, Dunn BL (2000) Deep-Ultraviolet Contact Photolithography. Microelectronic Engineering, vol 53, p 95

227. Ono M, Saito M, Yoshitomi T, Fiegna C, Ohguro T, Iwai H (1995) A 40 nm gate length n-MOSFET. IEEE Transactions on Electron Devices, vol 42, no 10, p 1822

228. Zell T (2000) Lithographie. Dresdner Sommerschule Mikroelektronik

229. Coopmans F, Roland B (1986) Desire: a novel dry developed resist system. Proc SPIE, vol 631, p 34

230. Seeger DE, La Tulipe DC, Kunz jr RR, Garza CM, Hanratty MA (1997) Thin-film imaging: Past, present, prognosis. IBM Journal of Research and Development, vol 41, no 1/2
http://www.research.ibm.com/journal/rd/411/seeger.html (Status of Nov. 2002), and
Goethals AM, van Den Hove L (2002) 0.18 μ Lithography Using 248 nm Deep UV and Top Surface Imaging
http://www.fabtech.org/features/lithography/articles/body4.169.php3   (Status of Nov. 2002)

231. Sandia National Laboratories (2002) News
http://www.ca.sandia.gov/news/source.NR.html (Status of Nov. 2002),
http://www.ca.sandia.gov/news/euvl/index.html (Status of Nov. 2002)

232. Fraunhofer Institut für Lasertechnik ILT (2002) Lampen für extremes Ultraviolett
http://www.fraunhofer.de/german/press/md/md2000/md12-2000_t3.html   (Status of Nov. 2002)

233. Muray LP, Lee KY, Spallas JP, Mankos M, Hsu Y, Gmur MR, Gross HS, Stebler CB, Chang TH (2000) Experimental evaluation of arrayed microcolumn lithography. Microelectronic Engineering, vol 53, p 271

234. Lucent Technologies (2002)
http://www.bell-labs.com/project/SCALPEL/ (Status of Nov. 2002)

235. Lucent Technologies (2002) Next Generation Lithography (NGL) Mask Formats
http://www.bell-labs.com/project/SCALPEL/maskformat.html   (Status of Nov. 2002)

236. Stickel W, Langner GO (1999) Prevail: Theory of the Proof-of-Concept Column

Electron Optics. J Vacuum Science & Technology B17, no 6, p 2847

237.   Kassing R, Käsmeier R, Rangelow IW (2000) Lithographie der nächsten Generation. Phys Blätt, vol 56, p 31

238.   Melngailis J (1993) Focused ion beam Lithography. Nucl Instr and Meth, vol B80/81, p 1271

239.   Miller T, Knoblauch A, Wilbertz C, Kalbitzer S (1995) Field-ion imaging of a tungsten supertip. Appl Phys, vol A61, p 99

240.   Prewett PD, Mair GL (1991) Focused Ion Beams from Liquid Metal Ion Sources. Research Studies Press Ltd, Taunton

241.   Shinada T, Ishikawa A, Hinoshita C, Koh M, Ohdomari I (2000) Reduction of Fluctuation in Semiconductor Conductivity by one-by-one Ion Implantation of Dopant Atoms. Jpn J Appl Phys, vol 39, p L265

242.   Wieck AD, Ploog K (1990) In-Plane-Gated Quantum Wire Transistor Fabricated with Directly Written Focused Ion Beams. Appl Phys Lett, vol 56, p 928

243.   Wieck AD, Ploog K (1992) High transconductance in-plane-gated transistors. Appl Phys Lett, vol 61, p 1048

244.   Bever T, Klizing KV, Wieck AD, Ploog K (1993) Velocity modulation in focused-ion-beam written in-plane-gate transistors. Appl Phys Lett, vol 63, p 642

245.   Hillmann M (2001) FIB-Lithographie. Dissertation Universität Bochum

246.   Fritz GS, Fresser HS, Prins FE, Kern DP (1999) Lateral pn-Junctions as a novel electron detector for microcolumn systems. J Vac Sci Technol, vol B17, p 2836

247.   Rogers JA, Meier M, Dodabalapur A (1998) Using printing and molding techniques to produce distributed feedback and Bragg reflector resonators for plastic lasers. Appl Phys Lett, vol 73, p 1766

248.   Becker H, Gärtner C (2000) Polymer microfabrication methods for microfluidic analytical applications. Electrophoresis, vol 21, p 12

249.   Scheer HC, Schulz H, Lyebyedyev D (2000) New directions in nanotechnology – Imprint Techniques. In: Pavesi L, Buzaneva E (eds) Frontiers of Nano-Optoelectronic Systems. Kluwer Academic Publishers, p 319

250.   Scheer HC, Schulz H, Hoffmann T, Sotomayor Torres CM (2001) Nanoimprint techniques. In: Nalwa HS (ed) Handbook of Thin Film Materials, vol 5. Academic Press, 1

251.   Chou SY, Krauss PR, Renstrom PJ (1995) Imprint of sub-25 nm vias and trenches in polymers. Appl Phys Lett, vol 67, p 3114

252.   Chou SY, Krauss PR, Zhang W, Guo L, Zhuang I (1997) Sub-10 nm lithography and applications. J Vac Sci Technol, vol B15, p 2897

253.   van Krevelen DW (1990) Properties of Polymers. Elsevier, Amsterdam

254.   Pfeiffer K, Fink M, Bleidiessel G, Grützner G, Schulz H, Scheer HC, Hoffmann T, Sotomayor Torres CM, Cardinaud C, Gaboriau F (2000) Novel linear and crosslinking polymers for nanoimprinting with high etch resistance. Microelectronic Engineering, vol 53, p 411

255.   Fa. Micro resist technology (2002)
       http://www.microresist.de/ (Status of Nov. 2002)

256.   Horstmann JT, Hilleringmann U, Goser KF (1998) Matching Analysis of Deposition Defined 50-nm MOSFETs. IEEE Trans, vol ED-45, p 299

257.   Schulz H, Lyebyedyev D, Scheer HC, Pfeiffer K, Bleidiessel G, Grützner G, Ahopelto J (2000) Master replication into thermosetting polymers for nanoimprinting. J Vac Sci Technol, vol B18, p 3582

258. Jaszewski RW, Schift H, Gobrecht J, Smith P (1998) Hot embossing in polymers as a direct way to pattern resist. Microelectronic Engineering, vol 41/42, p 575

259. Scheer HC, Schulz H, Hoffmann T, Sotomayor Torres CM (1998) Problems of the nanoimprinting technique for nanometer scale pattern definition. J Vac Sci Technol, vol B16, p 3917

260. Baraldi LG (1994) Heißprägen in Polymeren für die Herstellung integriert-optischer Systemkomponenten. Doktorarbeit an der ETH Zürich

261. Heyderman LJ, Schift H, David C, Gobrecht J, Schweizer T (2000) Flow behaviour of thin polymer films used for hot embossing lithography. Microelectronic Engineering, vol 54, p 229

262. Scheer HC, Schulz H (2001) A contribution to the flow behaviour of thin polymer films during hot embossing lithography. Microelectronic Engineering, vol 56, p 311

263. Srinivasan U, Houston MR, Howe RT, Maboudian R (1998) Alkyltrichlorosilare-based self-assembed monolayer films for stiction reduction in silicon micromechanics. J Microelectromechanical Systems, vol 7, p 252

264. Zimmer K, Otte L, Braun A, Rudschuck S, Friedrich H, Schulz H, Scheer HC, Hoffmann T, Sotomayor Torres CM, Mehnert R, Bigl F (1999) Fabrication of 3D micro- and nanostructures by replica molding and imprinting. Proc EUSPEN, vol 1, p 534

265. Heidari B, Maximov I, Montelius L (2000) Nanoimprint at the 6 inch wafer scale. J Vac Sci Technol, vol B18, p 3557

266. Roos N, Luxbacher T, Glinsner T, Pfeiffer K, Schulz H, Scheer HC (2001) Nanoimprint lithography with a commercial 4 inch bond system for hot embossing. Proc SPIE, vol 4343, p 427

267. Haisma J, Verheijen M, van der Heuvel K (1996) Mold-assisted nanolithography: A process for reliable pattern replication. J Vac Sci Technol, vol B14, p 4124

268. Colburn M, Johnson S, Stewart M, Damle S, Bailey T, Choi B, Wedlake M, Michaelson T, Sreenivasan SV, Ekerdt J, Wilson CG (1999) Step and flash imprint lithography: A new approach to high-resolution patterning. Proc SPIE, vol 3676, p 279

269. Colburn M, Grot A, Amitoso M, Choi BJ, Bailey T, Ekerdt J, Sreenivasan SV, Hollenhorst J, Wilson CG (2000) Step and flash imprint lithography for sub-100 nm patterning. SPIE Proc, vol 3997, p 453

270. Kumar A, Biebuck HA, Whitesides GM (1994) Patterning self-assembled monolayers: Applications in materials science. Langmuir, vol 10, p 1498

271. Xia Y, Zhao XM, Whitesides GM (1996) Pattern transfer: Self assembled monolayers as ultrathin resists. Microelectronic Engineering, vol 32, p 255

272. Xia Y, Mrksich M, Kim E, Whitesides GM (1996) Microcontact printing of octadecylsiloxane on the surface of silicon dioxide and its application in microfabrication. J Am Chem Soc, vol 117, p 9576

273. Xia Y, Qin D, Whitesides GM (1996) Microcontact printing with a cylindrical rolling stamp: A practical step toward automatic manufacturing of patterns with submicrometer sized features. Advanced Materials, vol 8, p 1015

274. Schmid H, Michel B (2000) Siloxane polymers for high-resolution, high-accuracy soft lithography. Macromolecules, vol 33, p 3042

275. Firma EVGroup, Austria
http://www.evgroup.com/ (Status of Nov. 2002)

Firma Obducat, Sweden
http://www.obducat.se/ (Status of Nov. 2002)

276.   Heidari B, Maximov I, Sarwe EL, Montelius L (1999) Large scale nanolithography using nanoimprint lithography. J Vac Sci Technol, vol B17, p 2261

277.   Haatainen T, Ahopelto J, Grützner G, Fink M, Pfeiffer K (2000) Step & stamp imprint lithography using a commercial flip chip bonder. SPIE Proc, vol 3997, p 874

278.   Montelius L, Heidari B, Graczyk M, Maximov I, Sarwe EL, Ling TG (2000) Nanoimprint and UV-lithography: Mix & match process for fabrication of interdigitated nanobiosensors. Microelectronic Engineering, vol 53, p 521

279.   Reuther F, Pfeiffer K, Fink M, Grützner G, Schulz H, Scheer HC (2001) Multistep profiles by mix and match of nanoimprint and UV lithography. Microelectronic Engineering, vol 57–58, p 381

280.   Xia Y, Whitesides GM (1998) Soft lithography. Angew Chem Int Ed, vol 37, p 550

281.   Eigler DM, Schweizer EK (1990) Positioning single atoms with a scanning tunneling microscope. Nature, vol 344, p 524

282.   Tan W, Kopelman R (2000) Nanoscopic Optical Sensors and Probes. In: Nalwa HS (ed) Handbook of Nanostructured Materials and Nanotechnology, vol 4, Academic Press, New York

283.   Betzig E, Trautmann JK (1992) Near-Field Optics: Microscopy, Spectroscopy, and Surface Modification Beyond the Diffraction Limit. Science, vol 257, p 189

284.   Trautman JK, Macklin JJ, Brus LE, Betzig E (1994) Near-field spectroscopy of single molecules at room temperature. Nature, vol 369, p 40

8   Extension of Conventional Devices by Nanotechniques

285.   Xu Q, Qian H, Yin H, Jia L, Ji H, Chen B, Zhu Y, Liu M, Han Z, Hu H, Qiu Y, Wu D (2001) The Investigation of Key Technologies for Sub-0.1μm CMOS Device Fabrication. IEEE Trans On ED, vol 48, p 1412–1420

286.   Wann C, Assaderaghi F, Shi L, Chan K, Cohen S, Hovel H, Jenkins K, Le Y, Sadana D, Viswanathan R, Wind S, Taur Y (1997) High-Performance 0.07-μm CMOS with 9.5-ps Gate Delay and 150 GHz fT. IEEE Elec Dev Lett, vol 18, p 625–627

287.   Iwai H, Momose HS. Ultra-thin gate oxides-performance and reliability.

288.   Ono M, Saito M, Yoshitomi T, Fiegna C, Ohguro T, Iwai H (1995) A 40-nm gate length n-MOSFET. IEEE Trans Elec Dev, vol 42, p 1822–1830

289.   Taur Y, Mil YJ, Frank DJ, Wong HS, Buchanan DA, Wind SJ, Rishton SA, Sai-Halasz GA, Nowak EJ (1995) CMOS scaling into the 21st century: 0.1 μm and beyond. IBM J Res Develop, vol 39, p 245

290.   Mikolajick T, Ryssel H (1993) Influence of Statistical Dopant Fluctuations on MOS Transistors with Deep Submicron Channel Lengths. Microelectronic Engineering vol 21, p 419

291.   Mikolajick T, Ryssel H (1996) Der Einfluß statistischer Dotierungsschwankungen auf die minimale Kanallänge von Kurzkanal-MOS-Transistoren. ITG-Fachbericht

138 „Mikroelektronik für die Informationstechnik", ISBN 3-8007-2171-6, p 183

292. Horstmann J, Hilleringmann U, Goser K (1998) Matching Analysis of Deposition Defined 50 nm MOSFETs. IEEE Transactions on Electron Devices, vol 45, p 299

293. Lakshmikumar KR, Hadaway RA, Copeland MA (1986) Characterization and Modeling of Mismatch in MOS Transistors for Precision Analog Design. IEEE Journal of Solid-State Circuits, vol SC-21, p 1057

294. Stolk PA, Schmitz J (1997) Fluctuations in Submicron CMOS Transistors. Proceedings of the Second Workshop on Innovative Circuits and Systems for Nano Electronics, Delft, The Netherlands, Sep 29–30, 1997, p 21

295. Wong HS, Taur Y (1993) Three-Dimensional "Atomistic" Simulation of Discrete Random Dopant Distribution Effects in Sub-0.1 μm MOSFETs. IEDM '93, Proceedings, Digest Technical Papers, Dec 5–8, 1993, p 705

296. Asenov A (1998) Random Dopant Threshold Voltage Fluctuations in 50 nm Epitaxial Channel MOSFETs: A 3D "Atomistic" Simulation Study. ESSDERC '98, Sep 8–10, 1998, Bordeaux, France, p 300

297. Skotnicki T (1996) Advanced Architectures for 0.18–0.12 μm CMOS Generations. Proceedings of the 26th European Solid State Device Research Conference ESSDERC'96, Sep 9–11, 1996, Bologna, Italy, p 505

298. Hellberg PE, Zhang SL, Petersson CS (1997) Work Function of Boron-Doped Polycrystalline Si_xGe_{1-x} Films. Letters, vol 18, p 456

299. Wirth G (1999) Mesoscopic phenomena in nanometer scale MOS devices. Dissertation, Universität Dortmund

300. Wirth G, Hilleringmann U, Horstmann JT, Goser KF (1999) Mesoscopic Transport Phenomena in Ultrashort Channel MOSFETs. Solid-State Electronics, vol 43, p 1245

301. Wirth G, Hilleringmann U, Horstmann JT, Goser K (1999) Negative Differential Resistance in Ultrashort Bulk MOSFETs. Proceedings of the 25th Annual Conference of the IEEE Industrial Electronics Society IECON '99, Nov 29–Dec 3, 1999, San Jose (CA), ISBN 0-7803-5735-3, p 29

302. Behammer D (1996) Niedertemperaturtechnologie zur Herstellung von skalierfähigen Si/SiGe/Si-Heterobipolartransistoren. Dissertation, Universität Bochum

9   Innovative Electronic Devices Based on Nanostructures

303. Sun JP, Haddad GI, Mazumder P, Schulman JN (1998) Resonant Tunneling Diodes: Models and Properties. Proceedings of the IEEE, vol 86, p 641

304. Haddad GI, Mazumder P (1997) Tunneling devices and applications in high functionality/speed digital circuits. Sol St Electron, vol 41, p 1515

305. Fay P, Lu J, Xu Y, Bernstein GH, Chow DH, Schulman JN (2002) Microwave Performance and Modeling of InAs/AlSb/GaSb Resonant Interband Tunneling Diodes, IEEE Trans Electron. Dev, vol 49, p 19

306. Chow DH, Dunlap HL, Williamson W, Enquist S, Gilbert BK, Subramaniam S, Lei PM, Bernstein GH (1996) InAs/AlSb/GaSb Resonant Interband Tunneling Diodes and Au-on-InAs/AlSb-Superlattice Schottky Diodes for Logic Circuits. IEEE Electron Dev Lett, vol 17, p 69

307.   Tsutsui M, Watanabe M, Asada M (1999) Resonant Tunneling Diodes in Si/CaF$_2$ Heterostructures Grown by Molecular Beam Epitaxy. Jpn J Appl Phys, vol 38, p L 920

308.   Ismail K, Meyerson BS, Wang PJ (1991) Electron resonant tunneling in Si/SiGe double barrier diodes. Appl Phys Lett, vol 59, p 973

309.   See P, Paul DJ, Holländer B, Mantl S, Zozoulenko I, Berggren KF (2001) High Performance Si/Si$_{1-x}$Ge$_x$ Resonant Tunneling Diodes. IEEE Electron Dev Lett, vol 22, p 182

310.   Ishikawa Y, Ishihara T, Iwasaki M, Tabe M (2001) Negative differential conductance due to resonant tunneling through SiO$_2$ / single crystalline-Si double barrier structure. Electron Lett, vol 37, p 1200

311.   Suemitsu T, Ishii T, Yokoyama H, Enoki T, Ishii Y, Tamamura T (1999) 30-nm-Gate InP-Based Lattice-Matched High Electron Mobility Transistors with 350 GHz Cutoff Frequency. Jpn J Appl Phys, vol 38, p L 154

312.   Zeuner M, Hackbarth T, Höck G, Behammer D, König U (1999) High-Frequency SiGe-n-MODFET for Microwave Applications. IEEE Microwave and Guided Wave Lett, vol 9, p 410

313.   Moon JS, Micovic M, Janke P, Hashimoto P, Wong WS, Widman RD, McCray L, Kurdoghlian A, Nguyen C (2001) GaN/AlGaN HEMTS operating at 20 GHz with continuous-wave power density >6 W / mm. Electron Lett, vol 37, p 528

314.   Eisele H, Haddad GI (1998) Two-Terminal Millimeter-Wave Sources. IEEE Trans Microwave Theory Tech, vol 46, p 739

315.   Peterson DF, Klemer DP (1989) Multiwatt IMPATT power amplification for EHF Applications. Microwave J, vol 32, p 107

316.   Eisele H (2002) High performance InP Gunn devices with 34 mW at 193 GHz. Electron Lett, vol 38, p 92

317.   Teng SJ, Goldwasser RE (1989) High-performance second-harmonic operation W-band Gunn devices. IEEE Electron Dev Lett, vol 10, p 412

318.   Brown ER, Söderström JR, Parker CD, Mahoney LJ, Molvar KM, McGill TC (1991) Oscillations up to 712 GHz in InAs/AlSb resonant tunneling diodes. Appl Phys Lett, vol 58, p 2291

319.   Mazumder P, Kulkarni S, Bhattacharya M, Sun JP, Haddad GI (1998) Digital Circuit Applications of Resonant Tunneling Devices. Proceedings of the IEEE, vol 86, p 664

320.   Otten W, Glösekötter P, Velling P, Brennemann A, Prost W, Goser KF, Tegude FJ (2001) InP-based monolithically integrated RTD/HBT MOBILE for logic circuits. Conf Proc of the 13th IRPM, Nara, p 232

321.   Kawano Y, Ohno, Y, Kishimoto, S, Maezawa K, Mizutani T (2002) 50 GHz frequency divider using resonant tunnelling chaos circuit. Electron Lett, vol 38, p 305

322.   Velling P, Janssen G, Auer U, Prost W, Tegude FJ (1998) NAND/NOR logic circuit using single InP based RTBT. El Lett, vol 34, p 2390

323.   Chen W, Rylyakov VP, Lukens JE, Likharev KK (1999) Rapid Single Flux Quantum T-Flip Flop Operating up to 770 GHz: IEEE Trans Appl Supercond, vol 9, p 3212

324.   Kholod AN, Liniger M, Zaslavsky A, d'Avitaya FA (2001) Cascaded resonant tunneling diode quantizer for analog-to-digital flash conversion. Appl Phys Lett, vol 79, p 129

325.   Sano K, Murata K, Otsuji T, Akeyoshi T, Shimizu N, Sano E (2001)An 80-Gb / s Optoelectronic Delayed Flip-Flop IC Using Resonant Tunneling Diodes and Uni-Traveling-Carrier Photodiode. IEEE Solid State Circ, vol 36, p 281

326.   Kawamura Y, Asai H, Matsuo S, Amano C (1992) InGaAs-InAlAs Multiple Quantum Well Optical Bistable Devices Using the Resonant Tunneling Effect. IEEE J Quant Electron, vol 28, p 308

327.   Goldstein S, Rosewater D (2002) Digital logic using molecular electronics. Proc of the IEEE Int Solid-State Circuits Conf (ISSCC), San Francisco, p 12.5

328.   Bryllert T, Borgstrom M, Sass T, Gustason B, Landin L, Wernersson LE, Seifert W, Samuelson L (2002) Designed emitter states in resonant tunneling through quantum dots. Appl Phys Lett, vol 80, p 2681

329.   Capasso F, Gmachl C, Paiella R, Tredicucci A, Hutchinson AL, Sivco DL, Baillargeon JN, Cho AY, Liu HC (2000) New Frontiers in Quantum Cascade Lasers and Applications. IEEE J Select Topics Quantum Electron, vol 6, p 931

330.   Liu HC (2000) Quantum Well Infrared Photodetector Physics and Novel Devices. In: Crandall BC (ed) Semiconductors and Semimetals. Academic Press, San Diego, vol 62, p 129

331.   O'Reilly E (1994) Quantum cascade laser has no role for holes. Physics World, p 24

332.   Faist J, Capasso F, Sivco DL, Hutchinson AL, Cho AY (1995) Vertical transition quantum cascade laser with Bragg confined excited state. Appl Phys Lett, vol 66, p 538

333.   Yang QK, Bradshaw JL, Bruno JD, Pham JT, Wortmann DE (2002) Mid-Infrared Type-II Interband Cascade Lasers. IEEE J Quantum Electron, vol 38, p 559

334.   Yang QK, Mann C, Fuchs F, Kiefer R, Köhler K, Rollbühler N, Schneider H, Wagner J (2002) Improvement of $\lambda = 5$ µm quantum cascade lasers by blocking barriers in the active regions. Appl Phys Lett, vol 80, p 2048

335.   Hofstetter D, Beck M, Aellen T, Faist J (2001) High-temperature operation of distributed feedback quantum-cascade lasers at 5.3 µm. Appl Phys Lett, vol 78, p 665

336.   Scamarcio G, Capasso F, Sirtori C, Faist J, Hutchinson AL, Sivco DL, Cho AY (1997) High-Power Infrared (8-Micrometer Wavelength) Superlattice Lasers, Science, vol 276, p 773

337.   Faist J, Müller A, Beck M, Hofstetter D, Blaser S, Oesterle U, Ilegems M (2000) A Quantum Cascade Laser Based on an n-i-p-i Superlattice. IEEE Phot Techn Lett, vol 12, p 263

338.   Page H, Kruck P, Barbieri S, Sirtori C, Stellmacher M, Nagle J (1999) High peak power (1.1 W) (Al)GaAs quantum cascade laser emitting at 9.7 µm. El Lett, vol 35, p 1848

339.   Hofstetter D, Faist J, Beck M, Müller A, Oesterle U (1999) Demonstration of high performance 10.16 µm quantum-cascade distributed feedback lasers fabricated without epitaxial regrowth. Appl Phys Lett, vol 75, p 665

340.   Tredicucci A, Capasso F, Gmachl C, Sivco DL, Hutchinson AL, Chu SN, Cho AY (2000) Continuous wave operation of long wavelength $\lambda = 11$ um) inter-miniband lasers. Electron Lett, vol 36, p 876

341.   Anders S, Schrenk W, Gornik E, Strasser G (2002) Room-temperature emission of GaAs/AlGaAs superlattice quantum-cascade lasers at 12.6 µm. Appl Phys Lett, vol 80, p 1864

342.   Rochat M, Hofstetter D, Beck M, Faist J (2001) Long wavelength ($\lambda = 16$µm), room-temperature, single-frequency quantum-cascade lasers based on a bound-to-continuum transition. Appl Phys Lett, vol 79, p 4271

343.   Colombelli R, Tredicucci A, Gmachl C, Capasso F, Sivco DL, Sergent AM, Hutchinson AL, Cho AY (2001) Continuous wave operation of $\lambda = 19$ µm surface-plasmon quantum cascade lasers. El Lett, vol 37, p 1023

344. Colombelli R, Capasso F, Gmachl C, Hutchinson AL, Sivco DL, Tredicucci A, Wanke MC, Sergent AM, Cho AY (2001) Far-infrared surface-plasmon quantum-cascade lasers at 21.5 μm and 24 μm wavelengths. Appl Phys Lett, vol 78, p 2620

345. Bewley WW, Lee H, Vurgaftman I, Menna RJ, Felix CL, Martinelli RU, Stokes DW, Garbuzov DZ, Meyer JR, Maiorov M, Conolly JC, Sugg AR, Olsen GH (2000) Continuous-wave operation of $\lambda = 3.25$ μm broadened-waveguide W quantum-well diode lasers up to $T = 195$ K. Appl Phys Lett, vol 76, p 256

346. Faist J, Gmachl C, Capasso F, Sirtori C, Sivco DL, Baillargeon JN, Cho AY (1997) Distributed feedback quantum cascade laser. Appl Phys Lett, vol 70, p 2670

347. Williams BS, Xu B, Hu Q, Melloch MR (1999) Narrow-linewidth terahertz intersubband emission from three-level systems. Appl Phys Lett, vol 75, p 2927

348. Köhler R, Tredicucci A, Beltram F, Beere HE, Linfield EH, Davies AG, Ritchie DA, Iotti RC, Rossi F (2002) Terahertz semiconductor-heterostructure laser. Nature, vol 417, p 156

349. Menon VM, Goodhue WD, Karakashian AS, Naweed A, Plant J, Ram-Mohan LR, Gatesman A, Badami V, Waldman J (2002) Dual-frequency quantum-cascade terahertz emitter. Appl Phys Lett, vol 80, p 2454

350. Nahata A, Yardley JT, Heinz TF (2000) Two-dimensional imaging of continuous-wave terahertz radiation using electro-optic detection. Appl Phys Lett, vol 81, p 963

351. Carr GL, Martin MC, McKinney WR, Jordan K, Nell GR, Williams GP (2002) High-power terahertz radiation from relativistic electrons. Nature, vol 420, p 153

352. Lynch SA, Bates R, Paul DJ, Norris DJ, Cullis AG, Ikonic Z, Kelsall RW, Harrison P, Arnone DD, Pidgeon CR (2002) Intersubband electroluminescence from Si/SiGe cascade emitters at terahertz frequencies. Appl Phys Lett, vol 81, p 1543

353. Soref RA, Friedman L, Sun G (1998) Silicon intersubband lasers. Superlattices and Microstructures, vol 23, p 427

354. Martini R, Bethea C, Capasso F, Gmachl C, Paiella R, Whittacker EA, Hwang HY, Sivco DL, Baillargeon JN, Cho AY (2002) Free-space optical transmission of multimedia satellite data streams using mid-infrared quantum cascade lasers. El Lett, vol 38, p 181

355. Martini R, Paiella R, Gmachl C, Capasso F, Whittacker EA, Liu HC, Hwang HY, Sivco DL, Baillargeon JN, Cho AY (2001) High-speed digital data transmission using mid-infrared quantum cascade lasers. El Lett, vol 37, p 1290

356. Blaser S, Hofstetter D, Beck M, Faist J (2001) Free-space optical data link using Peltier-cooled quantum cascade laser. Electron Lett, vol 37, p 778

357. Kosterev AA, Curl RF, Tittel FK, Gmachl C, Capasso F, Sivco DL, Baillargeon JN, Hutchinson AL, Cho AY (2000) Trace gas detection in ambient air with cw and pulsed QC lasers. Proc of CLEO 2000, p 513

358. Hofstetter D, Beck M, Faist J (2002) Quantum cascade laser structures as photodetectors. Appl Phys Lett, vol 81, p 2683

359. Takahashi Y, Fujiwara A, Ono Y, Murase K (2000) Silicon Single-Electron Devices and Their Applications. Proceedings 30th IEEE International Symposium on Multiple-Valued Logic 2000 (ISMVL 2000), p 411

360. Uchida K, Koga J, Ohba R, Toriumi A (2000) Room-Temperature Operation of Multifunctional Single-Electron Transistor Logic. Proc IEDM 2000, p 13.7.1

361. Likharev KK (1999) Single-Electron Devices and Their Applications. Proceedings of the IEEE, vol 87, p 606

362. Nakamura Y, Chen CD, Tsai JS (1996) 100-K operation of Al-based single-electron transistors. Japan J Appl Phys, vol 35, p L1465

363. Klein D, Roth R, Lim AKL, Alivisatos AP, McEuen P (1997) A single-electron transistor made from a cadmium selenide nanocrystal. Nature, vol 389, p 699

364. Ralph DC, Black CT, Tinkham M (1997) Gate-voltage studies of discrete electronic states in aluminum nanoparticles. Phys Rev Lett, vol 78, p 4087

365. Altmeyer S, Hamidi A, Spangenberg B, Kurz H (1997) 77 K single electron transistors fabricated with 0.1 μm technology. J Appl Phys, vol 81, p 8118

366. Matsumoto K, Ishii M, Segawa K, Oka Y, Vartanian BJ, Harris JS (1996) Room temperature operation of a single electron transistor made by the scanning tunneling microscope nanooxidation process for the TiO_x/Ti system. Appl Phys Lett, vol 68, p 34

367. Takahashi Y, Namatsu H, Kurihara K, Iwdate K, Nagase M, Murase K (1996) Size dependence of the characteristics of Si single-electron transistors on SIMOX substrates. IEEE Trans Electron Devices, vol 43, p 1213

368. Shirakashi J, Matsumoto K, Miura N, Konagai M (1998) Single-electron charging effects in Nb/Nb oxide-based single electron transistors at room temperature. Appl Phys Lett, vol 72, p 1893

369. Dolata R, Scherer H, Zorin AB, Niemeyer J (2002) Single electron transistors with high-quality superconducting niobium islands. Appl Phys Lett, vol 80, p 2776

370. Kawasaki K, Yamazaki D, Kinoshita A, Hirayama H, Tsutsui K, Aoyagi Y (2001) GaN quantum-dot formation by self-assembling droplet epitaxy and application to single-electron transistors. Appl Phys Lett, vol 79, p 2243

371. Graf H, Vancea J, Hoffman H (2002) Single-electron tunneling at room temperature in cobalt nanoparticles. Appl Phys Lett, vol 80, p 1264

372. Fu Y, Willander M, Wang TH (2002) Formation and charge control of a quantum dot by etched trenches and multiple gates. Appl Phys A, vol 74, p 741

373. Motohisa J, Nakajima F, Fukui T, Van der Wiel WG, Elzerman JM, De Franceschi S, Kouwenhoven LP (2002) Fabrication and low-temperature transport properties of selectively grown dual-gate single-electron transistors. Appl Phys Lett, vol 80, p 2797

374. Ono Y, Takahashi Y, Yamasaki K, Nagase M, Namatsu H, Kurihara K, Murase K (2000) Fabrication Method for IC-oriented Si Single-Electron Transistors. IEEE Trans Electron Devices, vol 47, p 147

375. Tachiki M, Seo H, Banno T, Sumikawa Y, Umezawa H, Kawarada H (2002) Fabrication of single-hole transistors on hydrogenated diamond surface using atomic force microscope. Appl Phys Lett, vol 81, p 2854

376. Pekola JK, Hirvi KP, Kauppinen JP, Paalanen MA (1994) Thermometry by arrays of tunnel junctions. Phys Rev Lett, vol 73, p 2903

377. Keller MW, Martinis JM, Zimmermann NM, Steinbach AH (1996) Accuracy of electron counting using a 7-junction electron pump. Appl Phys Lett, vol 69, p 1804

378. Krupenin VA, Presnov DE, Savvateev MN, Scherer H, Zorin AB, Niemeyer J (1998) Noise in Al Single Electron Transistors of stacked design. J Appl Phys, vol 84, p 3212

379. Zimmerman NM, Huber WH, Fujiwara A, Takahashi Y (2001) Excellent charge offset stability in a Si-biased single-electron tunneling transistor. Appl Phys Lett, vol 79, p 3188

380. Vettinger P, Cross G, Despont M, Drechsler U, Duerig U, Gotsmann B, Haeberler W, Lantz MA, Rothuizen HE, Stutz R, Binnig GK (2002) The "Millipede"-Nanotechnology Entering Data Storage. IEEE Trans Nanotechnology, vol 1, p 39

381. Inokawa H, Fujiwara A, Takahashi Y (2001) Multipeak negative-differential-resistance device by combining single-electron and metal–oxide–semiconductor transistors. Appl Phys Lett, vol 79, p 3818

382. Knobel R, Yung CS, Cleland AN (2001) Single-electron transistor as a radio-frequency mixer. Appl Phys Lett, vol 81, p 532

383. McEuen PL, Fuhrer MS, Park H (2002) Single-Walled Carbon Nanotube Electronics. IEEE Trans Nanotechnology, vol 1, p 78

384. Iijima S, Ichihashi T (1993) Single-shell carbon nanotubes of 1-nm diameter. Nature, vol 363, p 603

385. Bethune DS, Kiang CH, Devries MS, Gorman G, Savoy R, Vasquez J, Beyers R (1993) Cobalt-catalyzed groth of carbon nanotubes with single-atomic-layerwalls. Nature, vol 363, p 605

386. Thess A, Lee R, Nikolaev P, Dai H, Petit P, Robert J, Chunhui X, Young Hee L, Seong Gon K, Rinzler AG, Colbert DT, Scuseria GE, Tombnek D, Fischer JE, Smalley RE (1996) Crystalline ropes of metallic carbon nanotubes. Science, vol 273, p 483

387. Kong J, Soh HT, Cassell A, Quate CF, Dai H (1998) Synthesis of single single-walled carbon nanotubes on patterned silicon wafers. Nature, vol 395, p 878

388. Bachtold A, Fuhrer MS, Plyasunov S, Forero M, Anderson EH, Zettl ZA, McEuen PL (2000) Scanned probe microscopy of electronic transport in carbon nanotubes, Phys Rev Lett, vol 84, p 6082

389. Tans SJ, Verschueren ARM, Dekker C (1998) Room-temperature transistor based on a single carbon nanotube. Nature, vol 393, p 49

390. Guillorn MA, Hale MD, Merculov VI, Simpson ML, Eres GY, Cui H, Puretzky AA, Geohegan DB (2002) Operation of individual integrally gated carbon nanotube field emitter cells. Appl Phys Lett, vol 81, p 2860

391. Martel R, Schmidt T, Shea HR, Hertel T, Avouris P (1998) Single- and multi-wall carbon nanotube field-effect transistors. Appl Phys Lett, vol 73, p 2447

392. Franklin NR, Wang Q, Tombler TW, Javey A, Shim M, Dai H (2002) Integration of suspended carbon nanotube arrays into electronic devices and electromechanical systems. Appl Phys Lett, vol 81, p 913

393. Kong J, Cao J, Dai H (2002) Chemical profiling of single nanotubes: Intramolecular p-n-p junctionsand on-tube single-electron transistors. Appl Phys Lett, vol 80, p 73

394. Li J, Papadoupolos C, Xu JM (1999) Highly-ordered carbon nanotube arrays for electronics applications. Appl Phys Lett, vol 75, p 367

395. Choi WB, Chu JU, Jeong KS, Bae EJ, Lee JW, Kim JJ, Lee JO (2001) Ultrahigh-density nanotransistors by using selectively grown vertical carbon nanotubes. Appl Phys Lett, vol 79, p 3696

396. Bachtold A, Hadley P, Nakanishi T, Dekker C (2001) Logic Circuits with Carbon Nanotube Transistors. Science, vol 294, p 1317

397. Hu J, Ouyang M, Yang P, Lieber CM (1999) Controlled growth and electrical properties of heterojunctions of carbon nanotubes and silicon nanowires. Nature, vol 399, p 48

Index

Major references are given in **boldface** type.

W

X

Z